I0434523

THE GENE CONSPIRACY
or "Sorry...The Genes Made Me Do It"

THE GENE CONSPIRACY
or "Sorry...The Genes Made Me Do It"

Aris P. D'Avenal

Writers Club Press

San Jose New York Lincoln Shanghai

THE GENE CONSPIRACY or "Sorry...The Genes Made Me Do It"

All Rights Reserved © 2000 by Aris P. D' Avenal

No part of this book may be reproduced or transmitted in any form or by any means, graphic, electronic, or mechanical, including photocopying, recording, taping, or by any information storage retrieval system, without the permission in writing from the publisher.

Writers Club Press
an imprint of iUniverse.com, Inc.

For information address:
iUniverse.com, Inc.
5220 S 16th, Ste. 200
Lincoln, NE 68512
www.iuniverse.com

ISBN: 0-595-15131-0

Printed in the United States of America

CONTENTS

PREFACE

What I have done here is write a book about human behavior, under the assumption (and my firm conviction), that it is controlled by genes, or other chemical/electrical brain processes. Now I know you will object, and object vehemently. If true, it means that we have no will, no control and thus no responsibility. Well, I'll let you in on a little secret. If you think you are in control, it is because the genes…want you to think it. We all have a gene whose job is to make us think that we are the…masters of our domain, and to force us to object to suggestions that the genes…pull the strings. It is this gene that…compelled you to rush to the objection above. It is part of the…gene conspiracy that this book is devoted to uncovering.

But wait a minute, you will counter. How do you—the author—know this to be the case? Don't you also operate under the same gene's influence? Fair question, but I have an answer. The reason that I have been able to see through the plot and make these revelations, is because I am one of very few people in the world, with this gene defective—kaput. It's got to be defective, or it would have never allowed me to…think, let alone write, this book. As soon as the entire map of genes has been laid out—sometime in 2000-2010—I plan to have my genes checked to verify this suspicion. In the meantime, it is the only hypothesis that explains my ability to uncover the genes' nasty conspiracy, and my…unprecedented courage to write a book and expose it.

But what, pray, is the reason for all this, you will wonder. Why should our genes be involved in such a conspiracy against us? Aren't they there as part of our being? Isn't conspiring against us counterproductive—hurting their own cause? The answer might shock you. The genes are not there to serve us—they listen to a...different drum. They serve a higher, nobler goal—the preservation of our species. Oh, sure, in many cases this purpose coincides with our personal survival, but often it does not. When a conflict arises, the genes will abandon you with a vengeance—all in the name of their higher goal.

The book starts with several short stories of...weird human behavior. These stories are true, and they occurred in university towns, over a 40-year period. Intense, furious, discussions with my wife about these stories, lead to fierce disagreements on the origins of the behavior. The flagrant shortcomings of past and current psychological thought cry out for a new theory of human behavior, one more in tune with recent discoveries on the effects of the genes on the human condition.

The second part of the book outlines the proposed theory via a Socratic Symposium, with myself, my wife and my 3 sons as participants. Pointed discourses on the theory's applicability lead us to many subjects of current interest, including survival, happiness, religion, racism, youth violence, homosexuality, sexual and mental disorders, phobias—even people's interest in...celebrities. No subjects are left untouched, and no conclusions are avoided out of fear of political correctness. The discussions are passionate, poignant, and often...hilarious. The human condition is the most fascinating subject of all. I hope the book sheds some new light onto the various human quirks, follies and idiosyncrasies.

Aris P. D'Avenal

June 2000

1

The Little Episodes

The following real stories of singular human behavior took place over a forty-year period—early sixties to late nineties. Conventional wisdom has it that the...heroes of these stories are somehow in charge of their lives—responsible for the consequences. My refusal to accept this current-day psychological rationale has led me to the theory presented here—gradually, as the book evolves. This theory is sure to shake up most psychologists—real or...armchair types. It might even raise some rage against the...audacious author. But the duty is clearly laid out for me—I can not back away now, no matter what the consequences. Like the heroes of these stories, I am...forced to write this book—compelled to follow the dictates of my own genome, defective as it might be.

THE FLASHER...EARLY 60'S

The news was astonishing. A prominent professor at the University was arrested the previous night in the vicinity of the women's dorms. He had been knocking at coed windows...exhibiting himself. He was wearing a black overcoat, with nothing underneath it.

My wife Estelle brought me the news, early one evening. It seems that after his arrest, he ended up at the Psychiatric Clinic of the University Hospital, where Estelle worked as a nurse.

"He was highly agitated, almost in a panic when they brought him in," she said. "Apparently he had just realized what a terrible thing he had done. It was as if he had done it while in a daze—a...zombie like state. When he finally came to his senses, he went...berserk. The shock of the realization that he might have ruined his life was too much to bear, and he cracked. They had to sedate him to calm him down."

"That's amazing," I replied, though it took me some time to recover from the shock. I was a young man at the time. I knew very little about abnormal human behavior—the label given to such acts in the 60's.

"What in the world made him do it?" I asked puzzled. "What causes a man to risk so much just to get this kind of dubious sexual pleasure?"

"Who knows?" Estelle said pensively. "I suppose it has something to do with his childhood toilette training. Maybe his mother forced him to always cover his genitals when he...went, and this is his way of finally releasing himself from this terrible bondage to a...domineering mother." She said.

"Is that so? "I asked. "He actually flushed his career down the...toilette, because he wasn't properly...toilette trained?" I mused. "That is pretty ironic—very hard to believe."

And I didn't really believe it. But I wasn't about to argue the point with Estelle. I hadn't taken the courses in Psychology that Estelle had—I had no serious ammunition to challenge her judgment.

COPROLALIA[1]...LATE 60'S

"Boy, you look so depressed, I find it hard to greet you with the usual "how was your day Dear," Dear. "What happened?"

It was Estelle—my wife. We had been married for around 10 years now. I was just coming in, after having been privy to an incident that is hard to explain—let alone understand.

"I wish I knew what really happened," I said. "Let me just tell you what I heard at the rumor mill at the University, and then maybe you can tell me what happened."

"You know Dr G. the Professor in Chemical Engineering—they live down the street from here. He is a bit strange, with all those face ticks, the stammers, the hard swallows and all that, but he seems so nice—totally incapable of doing what they say he did."

"What did he do for heavens sake?" she was at the edge of her seat now.

"Well, it seems Dr. G. was meeting with the Dean of the College, and then all of a sudden, and with no provocation, Dr. G. started cursing at the Dean calling him a "son of .....", a "mother-.....," and some other pretty...colorful but rather nasty...epithets. Then, as if coming out of a trance—a...zombie-like state—he covered his face with his hands and stormed out of the room. Nobody has seen him since. There is a rumor that he checked himself in at the Psych Clinic of the Medical School—downtown."

"Hmm, " Estelle said thinking, trying to draw knowledge from the Psychology courses, from the back of her head. "He probably had an alcoholic, abusive father, who used this type of language daily when he was a child. That is why he has all those facial ticks and stuff."

---

[1] Coprolalia: A psychiatric term designating obsessive use of obscene language.

"What are you saying?" I replied. "Are you out of your mind? He went through a terrible, abusive childhood, with his father shouting obscenities at him all day, and after suffering all those years, he now decides to...emulate this kind of behavior? The result should have been the opposite—he should have turned himself into a gentle, soft spoken man." I was beginning to get argumentative by the late 60's—psychology courses not withstanding. I wasn't about to let Estelle get away with that kind of Psycho...mush, even though she had just completed her Bachelor's in Psychology and Nursing from a rather prestigious institution (Johns Hopkins).

"Like Father like Son, isn't that how the saying goes?" She replied. "If you are raised in such an environment, that is how you are going to end up—whether you want it or not," she replied. "We see examples of this everyday. But he'll be all right. A few...years in analysis, and the Doctors will bring these childhood traumas out in the open. Then he will become like you and me—I'm sure of it." She concluded.

"He won't" I thought, but didn't articulate this thought out loud. I was very skeptical of curing such ...mental singularities with years in analysis, but I didn't know why.

Dr G.—it turned out later—suffered from Tourette's Syndrome. The years in analysis appeared to have benefited only the analysts. It is now known, that the disorder is caused by a brain dysfunction.

The disorder ran in his family—on his mother side. It was *like Mother like Son* in his case.

A Gay Life Style?...Early 70's

"What a day!" I exclaimed, throwing myself down in my chair, exhausted. "I dare you to guess what happened at the department, today."

"What happened?" Estelle answered arranging herself nicely in a comfortable chair, ready to play audience in a Shakespearean play. She had grown to relish these small tragedies that played out occasionally in our microcosm of intellectuality—the University. She almost looked forward to some singular event, taking place.

"It seems that Dr. O. caught his wife in bed with her lover, a...woman. He left his house last night and is now living in his... office. Students reported seeing him during his office hours, filthy, unshaven—unable to concentrate and answer questions. The chairman had to intervene and ask him to move to...a hotel—he couldn't understand that the office was not his...home."

"It's bad enough to find out that your wife is unfaithful," Estelle volunteered, "but with a woman? That is a double whammy! I don't know what I would do if I found you in bed with a...man. I would certainly feel betrayed, but I would also suffer feelings of inadequacy—I'd be sure that I drove you to homosexuality," she sighed.

"That's pure rubbish," I protested (I picked up this expression in England—it makes you feel...aristocratic). "You can't drive someone to homosexuality. It's an inborn trait—you don't acquire it as you go along. His wife is either bisexual or Lesbian and he didn't know it. She may have just decided to become openly assertive about her identity—come out of the...closet, as the cliché goes."

"That's NOT what our church teaches," she replied vehemently. "Homosexuality is a choice you make, a choice to live an...alternate life style. I think their society is like a...cult—easy to get in, very tough to get

7

out. They...won't let you. But some day they will have to answer for it—it is a...sin."

"Aren't you a little...homophobic?" I asked. "This choice you are talking about. When and how do they make it? Is it like learning to drink...beer? Do you try the gay type of sex as a social thing—*it's the latest craze, you know*—and though at first it is...bitter, like beer, you continue doing it until it grows on you? Is that how people start on this alternative?"

"Yeah, something like that, but I am no expert in it—just stuff I heard at the...church socials," she confesses.

"Well, explain this to me, then," I asked. "Being gay is a stigma in our society. Most gay people stay in the closet—they hate to even tell their parents about it. They might lose their friends, their jobs—the agony is perpetual. Do you think that they...chose to continue to suffer these indignities, just to stay in the gay community—to keep doing something they didn't like in the first place?

"I guess, once you grow to like it, it becomes an...addiction," she replies. "It's like beer—your analogy, not mine. Can you stop drinking beer now that you grew to like it?" she asked, her body language indicating a small victory.

"If everyone in the society shunned me, if I was about to lose my friends, my parents, maybe even my job, not to mention YOU, I think I could manage to quit beer," I said.

"But let's not talk about me," I continued. "If they are addicted to gay sex, it's no longer a...choice, as you first claimed. And if so, it's not a...sin, either. You wouldn't send to hell all these alcoholics, or drug addicts would you?" I persisted.

"I certainly would," she said. "But not because they are addicted to gay sex, alcohol or drugs. I would do it because they *tried them* and allowed themselves to get addicted, even though they knew it was a...sin." She was strutting around the room on her high heels now—her victory was complete.

Hard to argue with such logic, I thought to myself. I better think some more about this issue.

LIKE FATHER, LIKE SON...LATE 70's

"I never would have suspected it," Estelle said coming in from a visit with her friend Betty, "but Gabriel is a wife-beater. He beats her up regularly. All these stories she tells people about her bruises and black eyes are flagrant lies—a cover-up. She finally confided in me tonight that she is thinking of leaving him—she's had enough. He nearly beat her to a pulp the other day—she spent two days in the hospital and not at her mother's place as she claimed."

"What! Gabriel?" I said, with a...Benny Hill look in my face. "Gabriel is the mildest, most compassionate fellow I know. Are you sure she is telling you the truth?"

"No question about it," Estelle said, "Betty was all black and blue, and there was no way she inflicted this kind of punishment on...herself—some bruises were in places she couldn't possibly reach. The man is a...beast—he ought to be incarcerated." She asserted.

"But what makes him abuse her like that?" I asked. "Is it something she does which infuriates him to that extend? Or was he...born a wife-beater, and can't stop himself from doing it?" I am always one to ask for the "why's" of a situation—a keen student of the human condition.

"The reasons men molest their wives are well known," Estelle answered. "Psychology has been studying such cases for decades, you know."

"What makes them do it?" I persisted. "Do they have the devil in them, and he makes them do it?"

"Your sarcasm is noted, but ignored" she said, taking on the airs of a University lecturer—just to irritate me, of course. "They do it, because their father abused their mother."

9

"Aha!" I interjected. "So you think that heredity played a role here? He inherited some tendency for...wife-abusing from his father?" I asked excitedly, thinking that she had started to think along my lines.

"Hereditary...wife-beating?" She started laughing—she almost fell out of the chair. "How do you come up with such...jewels?

She thought my suggestion...hilarious.

"If that's not it, was it, then, because he...enjoyed watching his father batter his mother? Did he see his father getting such a...kick out of it, that he decided to take it up as a...hobby?" I asked mockingly, annoyed that my idea was dismissed so hurriedly.

"Certainly not," she replied, "you shouldn't make jokes about such horrible acts. Gabriel, and all wife-beaters repeat what their fathers did, because that is what they saw when they were young and personable. They unconsciously think that the best way to treat their wife is like their father treated their mother, whom they loved dearly. You see that the environment in which you are raised plays a key role in wife-beating".

"But that makes no sense, at all," I protested. "If they spend their childhood watching their mother in daily agony and fear—shaken up and injured with regularity—wouldn't you think that they would develop an abhorrence for this behavior? Doesn't it make more sense that the son would swear not to repeat this dastardly deed himself? How can anyone theorize—and actually find people who adopt his theory—that seeing your own mother beaten to a pulp, leads you to do the same—all because of the influence of your childhood environment?

"I knew you would disagree with this explanation," she said. "Everyone knows that you are a...contrarian," Estelle retorted.

"You know what I think?" I said. "All the wife-molesters should line up all your psychologists and...beat them up to a pulp. Then, claim no responsibility—the environment did it."

THE I-5 BANDIT...EARLY 80'S

"Am I glad you are home early," Estelle said with apprehension in her voice. "The I-5 rapist is at it again, still on the loose. The last incidence just occurred a mile away from here, at a gas station. I am locking up the doors, and turning the lights out this evening—can't take any chances." She was visibly shaken—unusually glad to see me come home early that day.

We were in Oregon for the academic year, I was on a year's sabbatical from my permanent job. The rapist she was referring to was traveling up and down Oregon and California, doing his work within a few miles of both sides of Interstate 5.

"Oh, really?" I asked. "What is his latest escapade?" I wasn't too scared myself—he wasn't keen on raping any...men—so far, anyway.

"I just told you," she snapped. "He raped this girl—a cashier—at a station a mile from here. He dragged her to the bathroom, and...sodomized her. The girl reported that he was violent, sort of like in a trance—he wouldn't take no for an answer."

"What sort of sodomy was it?" I inquired, always willing to learn new techniques in sexual pleasures. "The word sodomy could describe any number of things, you know, though all are supposed to be...unnatural."

"I cannot believe you are so nonchalant about this—the rapist lurking at every turn. I don't know how he does it, though I suspect...*every which way*. Can we just not discuss the gory details?" Estelle snapped.

"All right," I said, "We can have this conversation after they catch him—you'll be calmer then. Now just tell me, why do you think the I-5 bandit goes around raping women? It obviously provides him with immense pleasure, or he wouldn't risk execution or spending the rest of his life in prison for it. Was it something he happened to come upon by...accident, or was he born with this obsession? What's the latest scuttlebutt at the...church socials on this mania?"

"Rape is a form of violence," Estelle said. "This is well known in Psychology—there is no need to refer to church social wisdom. The rapist is not getting sexual pleasure, just releasing pent-up anger. He does this because he hates his...mother, she was probably abusive or a...whore. By now he is so...confused, he thinks that all women are his...mother. This is his way of extracting his...revenge. It has something to do with the Oedipus complex as well, but I am not sure I can fully explain it."

"That's quite a theory," I said. "Abusive mother or maybe a prostitute, the Oedipus complex—the whole...catastrophe. How about a father who was not around or even sexually abused him, or that ubiquitous toilette training trauma?"

"Sure, go ahead and mock my explanations, but I don't see any profound analyses coming from your quarters. You have been having lots of laughs...ridiculing existing wisdom. Isn't it about time you came up with something to replace it? Or are you afraid of ridicule yourself?"

"I am working on it," I said. "I know that the rapist has something physically wrong with his brain. Some part of his brain dealing with sex, violence and/or control, is malfunctioning—he is not in charge there. That is why he is perceived as being in a trance, a...zombie like state. But I haven't totally formulated a general theory yet—nothing I can articulate at this point, anyway."

"But it won't be long before I do," I thought to myself. "Things were starting to jell—ideas were pouring in at a rapid pace. I just had to organize them into a unified theory, and then I am off to the races."

I'll Have Some...Vanilla...Late 80's

"I have had it with Antoine, I can't take it anymore," I said to Estelle, letting myself drop like a sack of potatoes on my favorite living room chair." I had just come home from a very hard day at the office—I had to unwind as soon as possible.

What did he do this time?" Estelle inquired. "I know you took him out to lunch to that Chinese restaurant. Did he pull the Chow Mein stunt again?" She asked smiling, eagerly awaiting the details.

"Bingo," I said. "I've never seen anyone so conservative—it drives me up a wall. I told him about all these great delicacies that Me Chen Lo—their Chinese chef—makes. The curies, the rice vermicelli Singapore, the Shrimp with ginger and scallions, the various sauces—you know. He listened carefully on the way there—seemed to take it all in. Well, do you know what happened after we picked up the menus?

"Let me just guess," she says, "He went through the menu with the waiter, asked every possible question about all the entries—the color of the sauce, the degree of spiciness, the nature of the oriental vegetables present—and at the end he ordered his usual...Chow Mein. Is that right".

"Bingo" I said, again. "Wouldn't that make you want to wring his neck?"

"No, not really," Estelle answered. "I know him well by now. He is an ultra conservative, in everything he does in life. Heck, it took him 20 years to replace his car, and then he did it...reluctantly. Have you noticed that he seems to wear the same shirt everyday?"

"Yes, I have—I hope he washes it at night!" I replied.

"Well, I have news for you. It is not the same shirt. It is many shirts that look the same, he buys them in six-packs like beer—his wife confided in me one day at the Mall. The man *hates change*."

"The funny thing is that he thinks he is a liberal, a...progressive—in his political views, anyway." I said. "How a guy who despises any sort of change in his life, can be such an advocate for change in...society is a big mystery to me," I said. "But even a bigger mystery is what makes him that way," I said.

"He's probably gone through some traumatic experiences as a child that turned him into this," she said. "Maybe his family moved abruptly to a new city and the loss of all his friends sent him to a mild depression. This could have caused failure in performance in the new school, no friends— misery all around. All that could be enough to turn him into a fanatic conservative—no moving, same food, even same...shirt," she theorized.

"I don't believe all that...bologna about childhood traumatic experiences causing changes in personality any more," I said. "All that is now...passé. The most likely explanation for his conservatism is a...gene," I asserted. "A gene of...conservatism that he probably inherited from his parents. His father looks exactly like a...conservative." I concluded.

"I expected that sooner or later you'd come up with some...nonsense like that," she replied. "I saw all those print-outs you've been reading from the Internet—all that misinformation about the human personality, and the effects of the genes. Just because it's in the Internet, it doesn't automatically follow that it's true, you know."

"If there is a gene for violence, for dyslexia, a gene for obsessive shopping, a gene for alcoholism or tendency to addiction and so on," I said, "I don't see what is so astonishing about a gene that causes conservatism. Or, even a gene for...lying, a gene for...jealousy, a gene for...fishing or hunting, a gene for raping or...wife-beating, a gene for...pederasty or bestiality, I can go on and on, you know. And let me tell you something..."

"Did I hear a knock at the door?" Estelle interrupted. She went over to open it.

"Oh, hi Antoine," I heard her say. "Come on in. Aris was just telling me about your delightful lunch. How about some ice cream? I just came back from the store where I picked some terrific new flavors—pecan fantasy, tropical audacity, and tri-flavor mambo. Or would you rather have some Black Forest cake with vanilla a la mode?" She asked him, as they came in towards the living room.

"I'll have the usual—plain…vanilla," he replied, as he sat down on the couch. "Two scoops if you don't mind."

Do You Have a Computer or Not?...Early 90's

"You know? I am sick and tired of this guy Omar—he is a bottomless pit of daily demands. He wants this and he wants that, his office isn't as good as he expected—the list is endless. I just about had it with him and ready to tell him so."

I was talking about Dr. Omar M., one of my faculty members at the Department. I had been the chairperson for over a year now, and I was trying to update all the office and Laboratory equipment—to bring it up to computer age standards. He was there at my office day in and day out, demanding a new Fax machine for his office, a new Scope for his lab—all the latest technology had to offer.

"Poor fellow was probably deprived when he was a youngster—it happens quite often," Estelle said, looking up from her knitting magazine. "Isn't he the one born in the Middle East?"

"Yes, he is," I replied, "but he was hardly deprived. His father was a Doctor. They were in the upper class of their society—certainly much less deprived than you or I. Stop using these silly psychological excuses for his behavior—it is caused by some defective genes, I'm sure," I snapped.

"There you go again, explaining everything using this mysterious...gene theory you are developing. But, if as you say, his hoggish behavior is due to some defective genes, why are you angry with him? Why don't you direct your anger at the...responsible genes?"

She said that laughing, pretending to still be reading a knitting magazine. But she wasn't fooling me—I knew how she loved this verbal...sparing with me, when I brought such problems home from the office.

"I can't be angry at his genes any more than I could be angry at his…deprived background. At least the genes live inside him, and he is…responsible for housing them there," I said.

"Anyway," I continued feigning lack of interest in continuing the dialogue, "I am going to wait and see what happens tomorrow, and then I will decide my next course of action."

"Why tomorrow? What happens tomorrow?" she asked nervously, careful not to show openly her excitement.

"I'll tell you, if you promise not to belittle my theories from now on," I said, knowing fully well that I could extract almost any promise from her, in exchange for some new juicy detail in the continuous saga of departmental machinations.

"OK, I promise," she said. "What happens tomorrow?"

"Well, I am expecting to get the faculty's written requests for the new up-to-date computers, donated by the M. and P. Computer company. Since there isn't enough of them to go around, I have already told them that requests are to be submitted only by professors who don't have any computer at all—either at their office or at their Lab."

"That sounds reasonable. But how is it related to Omar?"

"Omar has a computer, a very good one in fact—I gave it to him only a month ago. Yet, I saw the secretary typing up his request for one of the donated ones. I am quite anxious to see what he has written on it—what sort of twisted logic he has come up with, to justify asking for a new computer, when he just acquired one."

"Boy, that sounds quite exciting" Estelle replied. "You don't mind giving me a call when those requests reach your office—reading Omar's request to me, do you?" she asked.

"No problem," I replied. It was one of those small pleasures I could provide, with very little effort on my part.

Late the next afternoon, I called Estelle and read for her Omar's memorandum. It went like this:

To: Dr. A. P. D'Avenal, Chair of the Department
From: Dr. Omar M.
Subject: Request for a Computer

I am hereby expressing my immense need for one of the available computers. The situation in my Lab and Office is as follows:

Office: No computer: The newly acquired one has been moved to my laboratory, where it is badly needed.

Laboratory: No computer: The one that is presently there has been borrowed from my office.

....

Estelle was still laughing when I got home that evening—two whole hours after I called her.

A Racist...Gene?...Late 90's

"You read that story about that heinous crime in Colorado?" Estelle asked one evening, as we sat down at the verandah for our evening tea.

"It was a hate crime wasn't it? I asked. "Some white guy dragged a black man behind his pickup until he was dead. His whole body was in shreds—his face unrecognizable. They had to use his teeth to identify him, isn't that so?" I asked.

"How can anyone have such hatred in his heart for another human being, just because he...looks different?" She lamented. "How do you explain that with your new-found theory—the one that attributes most human traits to...genes."

"I have very little doubt that racism is caused by a...gene, and that all human being have it," I replied. In some people it is very strong—our recent murderer, for example—in others less so. In some it may be defective. But it is there—all the same. That is why it is so hard to...stamp out racism. It is everywhere, in every nation, in every tribe, in every race of the world."

"I find that very hard to believe," Estelle replied. "I have always thought that racism was...taught at the family or...social level—it is the product of the environment. Oh, I don't mean that someone sat us down and taught us to hate blacks or Jews, but we pick up on suggestions, small nuances—we get brainwashed in a subtle way. A racist...gene is totally new to me, and quite hard to swallow."

"You don't have to swallow it, it is already part of your constitutional essence," I replied.

"OK, I'll play along in this silly game just to see how far you are going with this bizarre theory," Estelle said. "I will agree that we all have this gene, for the time being, anyway. But why do we have it? Is it a kind of...disease, like polio?"

19

"I don't believe so," I said. "Diseases are caused by defective genes, or weak genes attacked by bacteria or viruses. Take Dyslexia for example, for which a gene has been recently identified. This gene—if defective—results in some damage in the brain, which causes dyslexia."

"Besides," I went on. "If racism was the result of some defective gene, it would show up only in a small percentage of the population. The fact that it is in all of us makes me conclude that it is there as a...normal gene—nature wants us to have it. In fact, I am compelled to conclude that those who are not racists are...abnormal—their racism gene is...defective. Another way to say it is that it is quite...natural to be racist, and unnatural not to be. The small suggestions, the little nuances, do help, of course—they strengthen the predisposition that is already there. But try to go against this gene—try to teach people not to be racist—and you got a huge problem on your hands."

"But why should nature put such a...nasty gene in our constitution? Why play such tricks on us, encouraging hate crimes, race massacres, ethnic cleansing, holocausts and the like? You've always claimed that nature wants us to perpetuate our species—that it helps us evolve to fulfill this goal. Isn't a racist gene...counterproductive? I think this gene theory of yours is leading you to...realm of the absurd!" She looked at me as if I was going... cuckoo.

"Your points are well taken?' I replied. "At first glance, the theory does appear to lead to some contradictions. But I still think it's right. Let us first look at this issue of whether nature put in us a...nasty gene or not. Putting it bluntly, I am proposing that we have a gene that makes us dislike and fear people that look different from us. This may appear loathsome at the present time—now that we live in multi-racial societies—but nature never figured that this would happen. Our genes were designed to help us live in a small tribe, and defend a territory for our survival. This depended heavily on our ability to recognize people from other tribes and chase them away, sometimes using violence to defend our borders. This...racist gene came in quite handy then—it aided the tribe in its survival. Even today, it probably helps nations fight a war. You need to hate the enemy to be…*effective*. And it is a lot easier to fight people that look different—Japanese for example during the 2nd World war—than it is to fight people that look similar—like the Germans, in the same war."

"So, this gene makes us hate those who look different from us?" Estelle asked.

"Primarily, yes," I said. "That is why it is easy for whites to hate blacks, blacks to hate Koreans, etc., and vice versa, of course. To accentuate the point and pacify the genes, we add fuel to the fire by using derogatory names for them, like the n-word for blacks."

"But why, then, do people discriminate against Jews or Arabs, even though they may look like them? Why do Catholics hate Protestants in Northern Island? We have many examples of racism, even when people look similar. The most recent one, I think, is in Yugoslavia, where hatreds run deep between peoples that look similar to each other. How does this gene of yours help explain these racist problems?"

"Fair question," I replied. "That is why I used the word "primarily" when you asked me about looks, above. The racist gene is there to help protect the tribe in general—not only the tribe's members and territory. It must also protect its culture, the part that binds the tribe together."

"Are you saying that some whites hate Jews because of their differences in culture?" Estelle asked quite puzzled.

"Precisely," I said. "An Arab does not hate a Jew at…first sight—they look quite similar," I replied. "Not until he knows that he is a…Jew. The same can be said about a Serb and an Albanian. In cases of similar looks, the gene is…activated when one finds out about the difference *in the culture*. It is quite hard for whites to hate Germans or Russians, and that is why *cultural* differences had to be emphasized, to help the gene do its… dirty work. So the Germans were baptized…Nazis, and the Russians… Communists, and the hatred flourished—the gene was content. Even so, the look-alikes do not suffer as much as the ones that are easily spotted as…different."

"Are you sure about that?" Estelle asked

"Well, There were no Germans put in concentration camps in the US during the 2nd World War, were they?" I replied. "The Japanese were, and the US Supreme Court even...approved it—that is how strong this gene is. If you look similar, you can... hide your cultural differences, and you will not be noticed. If you look different, you are... doomed, even if you fully adopt the culture."

"You see," I continued, "the gene reaches its pinnacle, when the other fellow differs in both looks and culture. A white Christian, for example, would find it quite easy to hate a black... Jew, easier that he would a white Jew or a black Christian."

"I am starting to think that this...racist gene might actually exist," Estelle said, and I nearly swallowed my...tongue from astonishment. It isn't often that I can get Estelle to agree to anything that deals with the...idiosyncrasies of human behavior.

"Still, I find it hard to admit that we have no choice in the matter," she continued. "Even if the gene is there, encouraging us to be racists, why can't we fight it off using our brain? The gene is not holding a gun, demanding that we be racist or else?" she insisted.

"Well, not exactly," I replied. "The gene—like all behavioral genes—is much ...cleverer than that! It does its work subtly. It effects your brain's thinking, and makes you think that what you are doing is...logically deduced. It provides you with all kinds of... excuses to be racist—it convinces you that you are justified in being so. But your overall behavior is also effected by its...strength, and by whether it is...energized or not. It probably also works in...unison with other genes of similar activity (like the...survival gene, for instance), so the issue is much more complex than that—we need to wait for science to uncover all these genes to fully understand it."

"You are starting to lose me now," Estelle interrupted, " you are bringing up too many new issues. I can guess that the strength of the gene would determine the extent of our racism, and thus our ability to think independently of the gene. Is that what you are saying?" Estelle wondered.

"Exactly," I replied. "Some people have such a weak racist gene that their brains are totally unaffected. They are the ones that champion the fights against racism, befriend and even marry people of another race. Others, have the gene in…medium strength—they are the ones that join in these fights, especially if the racism showed up…somewhere else. Then, you have the ones with such a potent gene, they are compelled to commit repulsive crimes. I presume that this is the case with our present murderer in Colorado—though his crime could be due to a defective gene as well."

"OK, fine, that is exactly what I thought anyway—you don't need to belabor the obvious," Estelle replied. "But what was all that about the gene being energized. Are you saying that sometimes the gene sits dormant and lets your brain…think for itself?"

"Quite so," I replied, "If there is no perceived threat to a person or his tribe, the gene lies dormant and your thinking is clear—you fancy yourself being tolerant."

"Do we have some examples of that?" Estelle asked.

"Yes, we do," I replied. "Most people are actually amazed that others— somewhere else—are racists. So you had the northerners in the US criticizing the southerners for discrimination against the Blacks, or the French and the English accusing the Americans of racism—the accusers were thinking logically because there was no perceived threat in their…neck of the woods. The picture, of course, changed dramatically when Blacks moved to the North and other races moved into England and France. As soon as the people perceived a threat, the gene was awakened from its lethargy, and logic suffered a serious setback. This gene is quite sneaky—it can…fool pretty much everyone. Of course, even when logic is gone, the person is unaware of it—it is all part of the…gene conspiracy"

"Gene conspiracy? What in heaven's name are you talking about now?" Estelle looked surprised.

"There is a conspiracy all right, some collusion between all the genes and...Nature, but I haven't totally figured out the how or why, so I'd rather not talk about it yet," I replied.

"Well, conspiracy or not, you paint a very bleak picture of the human character," Estelle remarked. "If this gene really exists and, mind you, I am not totally converted yet, what hope is there for the future of humankind? How are we going to be able to live peacefully in these multiracial societies that we are creating?"

"I don't know," I replied, "I don't pretend to be a social philosopher, I haven't really thought about the solution—it's hard enough making the case for the gene's existence. But if it does exist, science will find it, and the theory will be vindicated. Then the first thing that people must do is admit that they are all...racists. At present, everyone accuses the other of racism, and nobody admits it. You can't make much progress that way."

"Well, then," Estelle said, "since you are pretty convinced of your theory's validity, let me hear you say it out loud that you are also a...racist!"

"Who, me... a racist?" I replied "That is ridiculous! I married you, didn't I? According to the theory, people who marry across racial lines have a very weak racist gene, remember?"

For the sake of the reader, and in case I haven't mentioned, Estelle is half Filipino.

"I remember just fine," she replied. "I also remember you saying that having the gene in medium strength is...normal! Are you admitting then that you are...abnormal?" She wouldn't let go—she had me...cornered.

"No, I am not!" I protested. "Sure I married you, but you are not entirely of a different race—you are half Irish," I said, throwing up enough...smoke to allow me to repel the accusations.

You see, my racist gene gets...upset, if I am called a racist—it's somehow all part of the...gene conspiracy.

Estelle remained silent, trying to sort out the nonsense I threw at her, and I left the room. It was one of the ways we had silently agreed to end our arguments.

2

The Sympsium

The dinner Estelle prepared was exquisite. Stuffed mushrooms, potato croquettes and casseri cheese pies for appetizers, grilled pike with potatoes au-gratin and tabouli salad for the main course, and crunchy baklava for dessert—only Estelle knows how to make it that crunchy.

"So, what is the occasion for this phantasmagoric feast?" asked Christopher, the middle son, as we moved toward the living room for cognac and liqueurs.

"Can't you guess?" Said Andrew the first son, aspirant to be...head of the family. "Why do you think he sent us those little...episodes to read—the ones that span almost four decades?"

"I need you all here to help me finish this book I am writing," I said. "I need to execute a Socratic style discussion—to bring out all points of view. You know that I abhor writing something in the lecture style—it is too monotonous. Those little episodes are meant to introduce the subject, to prick your interest, to cause some...stirrings in your brain."

"Come on, dad," said Paul, the third son. "Those stories were actually the...homework for tonight's discussion, right?

"Actually, I got to admit that I enjoyed reading those stories—they brought back some memories" Andrew volunteered. "The one about the

I-5 bandit in Oregon—I recall that quite well. Corvallis was in a state of panic for more than two weeks if I remember right."

I should mention—parenthetically—that all three of our sons are electrical engineers—all chips off the old block, profession-wise.

Now, you ask, what could three electrical Engineers and one nurse—don't forget Estelle—contribute to a socio-psycho-philosophical discussion about genes and the human condition?

Well, I asked them to participate because their totality combines for a respectable amount of intellectual punch (2 Ph.D's, one Master's, one Bachelor's), and all of them are opinionated, argumentative, and...stubborn. Genes for all these three character traits can be found in both their parents—but that is another matter. I also needed to present my ideas to typical readers of the book—educated, independent and...busy—so I can see their initial reactions to its revolutionary ideas. I knew that most of them will attack my theory—their genes will not appreciate what I have to say. But I was resigned to this opposition—I actually welcomed it. I wanted to observe how their genes will react when I started to expose their conspiracy—I was sure I could use these observations to my advantage.

"Personally, I liked the one about the prof who suffered from Tourette's syndrome," said Christopher. "I actually know a guy who can't say anything unless he uses the "f" word in between every other...."

"You aren't really going to discuss your "gene" theory when there is the Wimbledon final on in 15 minutes," Paul interrupted. "Sampras is playing Agassi—it is the game of the century. If Sampras wins, he ties the all-time record for grand-slam event victories, that Emerson held for...

"I knew something like that might happen so I am not taking any chances—I am taping it," said Christopher. "But I might leave early anyway, I pretty much agree with Dad, so I am not really needed—I am sure he can hold his own in this discussion."

I cursed myself, silently, for making the mistake of setting up this discussion during the Wimbledon final—I wouldn't mind watching it myself. But I had to go on, anyway. It's very hard to get all four of us together—they are very busy people.

"I'll watch it and come in now and then to bring you up to date, if you want" Estelle said "I certainly don't have to be here in this discussion. Your father and I argue perpetually about his…genes, his…godlessness, and all these other…apocalyptic views he brings to the table nightly—especially since he retired. Some day he will have to answer though, and…

"OK, enough of this…buffoonery" I said emphatically. "The issue is very important—we'll be discussing people's personality here."

"All right," said Andrew. "Go ahead and present your theory. The sooner we get through the…boring technical details, the better."

"And make it snappy," Paul added. "Maybe we can still make the game!"

"As an introduction," I started, "Let me recall that what we have so far, is the *nature vs. nurture* debate—a bunch of scientists arguing incessantly about the relative effects of heredity and environment. This theory, in my opinion has become a…joke—they are now arguing about the…percentages. Heck, in the last study that Andrew sent me, the guy was trying to prove that heredity was 47.5% and not 50% as the …opposing school was claiming."

"Did you also notice the confusion about what the environment is?" Andrew volunteered.

"I most certainly did," I replied. "It used to be that the word environment represented only socio-psychological factors—the lines were clearer. Hereditary traits were what you were born with, and environmental traits what you acquired as you grew up—in a given socio-psychological milieu. Take, for example, the trait of lying. Most scientists used to think that you were born either neutral (tabula rasa) or with an inborn tendency to tell the truth. The environment, however, quickly turned you into a liar. In most—normal–cases, you learned how to tell small, white lies. If, however, you lived with people who were pathological liars, you unavoidably became one, too."

"That's a crock of s..." Christopher said. "George C., down the street here is a chronic and habitual liar, and both his parents are...god-fearing people. I think he was born that way—with a strong gene for lying."

"Recently," I continued, purposely ignoring his remark "the phrase environmental factors has been expanded to include physiological factors, as well—radiation, cigarette smoke, and the like—we now know that they can cause serious effects, both mental and physical. But if that is the case, when does the so-called "environment" come into effect? After you are born, the minute you pop out? Or is it the moment you are...conceived? A mother's womb, after all, is an environment, a very pleasant one in most cases, people say. But it is an environment that can be subjected to physiological effects, like radiation, for example. In fact, late research, as we speak, is finding that the period we spend in the womb is very important in shaping a person's future—physically or mentally."

"Oh, really?" Paul asked. "How so?" His wife was pregnant –the issue perked his interest.

"Research is showing[2] that besides the usual stuff like mother's smoking, drinking, and diet, other factors can effect your genes in the womb—alter them, kill them, or even "reprogram" them so they do not cause the body to do what was originally intended. One study showed if the mother is under heavy stress, stress hormones can flood the womb and alter the fetus's gene that controls stress response. The offspring may not be able to handle stress, later in life. In another study, diabetic mothers flooded the womb with glucose, and the offspring eventually developed diabetes. I think the...inevitable conclusion is that you may be born with traits that are stamped inborn or hereditary, but which, in fact, had been altered by the environment in the womb and thus were not hereditary. To use their terminology, you were born with natural traits that were actually...nurtured. I told you the whole debate has turned into a...fiasco."

---

[2] See "Shaped by Life in the Womb", Newsweek, Sept, 27, 1999

"Personally, I think the effects of the environment start even before… conception," said Andrew. "Everyone knows by now that drugs, alcohol, smoking, etc. effect men's sperms and women's eggs. Heck, if the woman went and had an X-ray the day before conception, her egg may have been altered and then…

"I totally agree," interjected Christopher. "Shelby, the guy at work was told by his pathologist that the reason for his low sperm count was…his hobby—bicycling. It seems that the chronic rubbing of his testicles on the bicycle seat increased their temperature—killed most of the sperms. It might have even altered the chromosomes, or whatever, on the surviving sperms, and if he had any children, they may have been born with three…ears. The doc told him to change his hobby to…backgammon."

"This whole question on when the environment starts reminds me of the abortion rights vs. the right to life debate," Paul said. "Maybe we can compromise and agree that the environment starts on the womb… three months after conception—I am an advocate of abortion rights, myself. Anyway, the game starts in five minutes."

"Science is not…politics and compromises like that are unscientific," I replied with a deadly look on my face—expressing displeasure at his continued…tennis interest. I can also…act—if necessary.

"Let me have a word here," Andrew interrupts. "OK, I think you just showed that if the…ubiquitous physical factors are included in the nurture part, the nurturists have just about …won the debate. This gene theory of yours—you are not trying to prop up the… naturists with it, are you? People usually associate the word genes with…heredity."

"Not at all, not at all," I replied. "I told you the debate is…irrelevant as far as I can see. When I say that a person's personality traits are traced to…genes, I don't mean to genes he…inherited—we are all born with the same types of genes, anyway. What makes us different, gives us our special character, is the… strength of each gene—a notion we will discuss extensively after these preliminaries. This strength could be inherited—it often is—at the point of conception, anyway. But it is a long ways from…conception to adulthood.

All kinds of physical forces could...attack these genes and change their...strength—even render them defective. It's not an argument that aids the naturists, I assure you. But some of these questions will be answered by the theory—if you ever let me start explaining it to you" I protested.

"The way I see it," Christopher interjects, "it's like a stagecoach in the old...westerns. It never ends up, the way it started out. Many forces attack it, as it gallops through the new territories—Indians, bandits, tornadoes, stampedes and so on. Even fighting among the passengers. Maybe all these genes fight with each other, as well—this could alter their effectiveness also," he added. Christopher is the...flick buff in the family—never leaves a movie...unseen.

"I'm sure they fight—thanks for bringing this up," I replied. "When we agonize about a course of action, we are probably feeling the effect of a gene...conflict" I added.

"I noticed you stressed the words physical forces in your explanation," Andrew returned. "Are you saying that the values, the strength of these genes can not change by psychological or sociological forces?"

"I think that such influences are minimal," I replied. "But we can talk about it later—you might be able to change my mind. I'll come up pretty forcefully when we get into the details, I am sure. If it doesn't, I promise to take it up in a special section with Frequently Asked Questions (FAQ's)—it is quite fashionable these days. Actually, I think I will baptize the section Rarely Asked Questions (RAQ's)—I don't imagine many people...itching to ask them."

"OK, then," Paul retorted, "You tell us your gene theory and let us attack it to see how well it takes the heat—how...pertinent it remains at the end. I think I am ready for it right now! Just make it snappy—we may still have a chance at the tennis match."

"That's my boy," I said. "That is all I have been waiting to hear. We got through the most boring stuff now—the rest is subject to...objections and therefore...fun."

I then stood up, assumed the expression of the professorial… know-it-all, and started on my long awaited…diatribe.

GENES AND THE SURVIVAL PRINCIPLE (GSP)

"First of all," I said, "we are all familiar with the fact that right after conception, the fetus has its DNA with the usual helix, and around 100,000-140,000 genes in it—most haven't yet been deciphered by scientists. One group of these genes deals with the physical growth of the fetus and eventually the human being. There are genes which ensure that we have one head rather than two, that our...rectum is not near our nose, that the proper number of all the other extremities and organs are there—all functioning properly. There are also genes that...manage the other genes—ensuring that all tasks are carried out correctly and at the proper time. The whole gene structure is probably like a huge company with the upper management, the white-collar workers, the blue-collar workers, etc. This much is pretty well accepted by the scientific community with no serious disagreements."

"Can I say something—it will only take a second," Christopher said. "Have they yet found the gene that is responsible for the number of sexual organs that we have? This one guy I know has two, and he claims...."

"We can all guess what he claims," Andrew interjected. "Even if they found the gene, I seriously doubt that doctors will try to alter it so that people can have more than one—there is ample trouble around with just the one, anyway."

"Please, let us have no more interruptions for a few minutes till I lay out the groundwork," I said. "After that, I welcome the interruptions—that is the point of this happening."

"My contention is two-fold. First, that there is also a second group of genes, responsible for our...personalities—all of our character traits, good or bad. And second, that the various traits we have, are placed there for one and only purpose; to help the species survive, evolve and flourish.

Unless, of course, they come from a…defective gene or genes, in which case they are diseases or abnormalities. The theory—we will call it Genes and the Survival Principle, (GSP for short)—is quite simple. Yet, it will turn out to be very controversial—you will see as we develop it."

"Are you really going to explain the…abyss of human behavior, using just this GSP?" Andrew wondered. "This, I got to see."

"I am going to give it a try—hope to have fun doing it—with your help, naturally," I replied. "If I can explain 90% of human behavior this way—I figure I'm home. The other 10% may be due to my…personal lack of intelligence to do so. Other thinkers may be able to do it later—using the same theory, of course. Besides, when it comes to the human character, I don't know of any theory that has explained everything, to everyone's satisfaction, at this point in time."

"OK," I continued, pulling out my…only chart. It looked like this!

"Boy " said Christopher, "Dad is sparing no expense, he's got a multi-media presentation here."

"Never mind the wisecracks." I interrupted. "This chart is a rectangularized form of the DNA Helix. Every rectangle represents a gene. The DNA of every human being has some tens of thousands of these—all assigned a specific task that causes your physical stature and your personality[3]. The issue is, of course, very complex so I have simplified it here in

---

[3] The genes are actually blueprints (coded directions) for making proteins that do the jobs, but our discussion here is simplified.

my own way. I have decided to assign to most genes a...strength—I like to think of it as a value in the scale from 0 to 10."

"Oh, yes the gene strength—I remember you peddling it a couple of years ago. Is that part of this eerie...gene conspiracy, you are uncovering? By the way, Sampras took the first set quite handily."

"Let me explain further," I replied, ignoring the wisecrack about the conspiracy. "We may not need strengths for genes that do some...physical task like making insulin, for example, though they may have them just as much as any gene. But we do need strengths for personality type genes, to explain the variations in character traits. Take the gene of...conscience—I think there is definitely one of those in case you are wondering—even though it hasn't been found yet. A very conscientious person has a gene of value 9 or 10. The so-called iceman (no sense of right and wrong) has a gene with strength 0 or 1—it might even be that his gene is defective or plain...dead[4]. A normal human being has a gene of value from 4.5 to 5.5—most of the time he behaves—but once in a while, he pulls something unconscionable."

"Another way to think of it," Andrew said, "is to tie it in, with their turn on switch. Some stimulus turns the gene on, as we know. The amount of the stimulus needed is Dad's strength here. In the iceman, no stimulus can turn the switch on—the switch is broken. In the very conscientious fellow pretty much everything turns it on, or it may also be broken—permanently open. I've been reading up on genes lately, just to keep up with Dad's arguments here."

"Hey, that would be a good way to explain the various types of sexual attractions," Christopher said. "If there is a gene that decides what you are, then the 0 would be the homosexual, 5 the bisexual and 10 the heterosexual.

---

[4] Most genes have a *turn on* and *turn off* mechanism. If the *turn on* mechanism is damaged, the gene is as good as...dead. Some genes are on all the time, like the gene that keeps the code for making hemoglobin

In between you have the various other shades we are often amazed at—only God knows what a 2.7 will get you," he added.

"Well," Andrew comes in again, "you could explain that by saying there are two genes of sexual attraction to another human being—one for attraction to males another for attraction to females. All of us get both these genes. Males have only the female attraction gene openable, and the male attraction gene permanently closed, and vice versa for the females. A homosexual then could be a person where these are mixed up."

"What about the bisexual? Asks Christopher.

"Well maybe both switches can open, depending on the stimulus." Andrew says.

"I think both explanations are possible, and others as well." I said. "The key thing in the theory is not how exactly the genes determine your…sexual preference, but that *they do*. Most people will disagree with this contention—they are within their rights, of course. But controversy aside, the genes will be assumed to have strengths from now on, and this concept will prove to be very useful in the sequel."

"But how do the genes come to have different strengths?" Estelle asked. "Not that I agree with your theory, mind you," she added

"Many ways, I suppose," I replied. "But you are forcing me to become more…technical and I don't like it—it is boring and outside the…scope of the book, as they say. The health of their constituent parts[5] depends on many things: heredity, radiation or toxic chemicals, bacteria, viruses, diet, etc., both inside the womb, and during one's lifetime. All these factors can alter these parts somewhat and cause the various strengths in question. Maybe it is also nature that randomly throws out various strengths in these genes to create all these personalities, as perpetual…mutations. I

---

[5] Genes are made up of small pieces called *nucleotides*—The DNA has over 2.5 billion of them.

don't really know at this stage—researchers are going after the answer as we speak, I suppose."

"I told you it was like the old...stagecoach running through the West, being attacked...

"By the way," I continued, interrupting Christopher, "the following is a key point, so please pay special attention to it. All character traits, mental disorders, etc. will be attributed to some...gene in this discussion—the gene that controls the part of the brain these traits originated in. Changes in this gene produce changes in the trait. However, I think it possible to have that same part of the brain directly hit by some physical force (chemical, electrical, etc.), and this could also alter the trait. In addition, we must remember that genes only carry blueprints for the manufacture of proteins that do the actual tasks. Even small changes in the gene constituents can change the amount of the protein produced and thus change the effect of the gene. It is easy to see that the complexities are immense. In any event, and since we don't know what really happens, we will *blame* the gene here—it will make our discussion easier to follow.

"So you plan to contend that personality traits are due to genes, and their strengths" Andrew said. "Your key point is that they are not caused by socio-psychological factors, just physical ones, right?

"Right," I replied.

"OK, I am starting to understand the issue a little better now," Christopher said. "Let's take the gene of...racism—you talked about it extensively in the short story before our symposium here. I guess the white supremacists have this gene with strength 10, or even defective. But you also talked about...dormancy or lethargy there, if I am not mistaken. Something about the gene...waking up when the person perceives a threat, and... sleeping the rest of the time. Does this dormancy appear in all genes?"

"I am glad you brought that up just now," I said, "I can expand on it at this point quite nicely. Most genes have an "on" and an "off" mechanism, as it has already been mentioned—very few are always on. When they are off, they are dormant. Take the gene of…religiosity, for example. We all have it, most people with strength 5—myself sitting at 0. The ones with strength 10 are perpetual slaves to this gene, they think of God or some religion, praying every minute throughout the day. In others, the gene's off mechanism went…kaput—again zero dormancy. For most people, though, the gene is usually dormant. They don't think much about the Almighty until a catastrophe comes around. Then they drop straight down on their knees—the gene wide awake with its demands."

"I know exactly what Dad means," said Paul. "Plenty of athletes have— or pretend to have—a high valued religiosity gene. They irritate me when I hear them talk during interviews. They prayed to God for a victory before the game—they thank God for the victory afterwards. Heck, when did God become a football or basketball fan, and a…partial one at that?"

"I'm with you there," I said. "Invoking the name of God at every turn actually trivializes prayer in my opinion."

"What about fear?" Christopher asks. "That is got to be a gene that is also dormant most of the time."

"Yes it is—in normal living, of course, though it is possible to conceive of situations when you live under constant fear—gene always on," I replied.

"What about strength value of the fear gene?" Christopher wondered. "Could we safely say that a guy with gene strength 10 is…soiling his under-shorts even at the sight of minimal threats to his life?"

"Yeah, sure," I replied. "Soiling yourself is the last desperate act to…repel the enemy—the use of disgusting visual effects and foul odors. If the gene strength is very low, the act plays out even at the sight of quite minor dangers."

"But before we get to any more details," I continued, "we need to put some order in this whole issue. Here's a list of genes that I have made, grouping them into some major but arbitrary categories. These are strictly my own—nothing sacred about them."

At this point I passed out my second article of multimedia presentation—copies of a list of gene groupings I had made, for their handy use during the discussion. It looked like this:

The Talent Group

Singing, dancing, music composing or playing, specific instrument playing (piano, drums, etc), painting, sculpturing, writing (prose, poems, plays), speaking (gift of gab), athleticism (football, basketball, tennis, track, circus, etc.), mathematical thinking, inventing, foreign language learning, etc.

The Survival/Perpetuation Group
(Self and Tribe)

<u>Life and Territory</u>: Hunger, thirst, hunting, fishing, gathering, stockpiling, self-defense, survival, fear, violence, aggression, hate, killing, cursing, racism, patriotism, bravery, heroism, altruism, machinating, climbing in hierarchy, interest in politics, aggressive sports, etc.

<u>Love and Culture</u>: Love (heterosexual, mother-child, family, friendship, etc.), sexual attraction, procreation, parenting, grand-parenting, language, communicating, socializing, religiosity, moralizing, rituals, singing, dancing, performing, writing, non-aggressive sports, philanthropy, gossiping, interest in celebrities.

The Malfunction Group

<u>The Phobias</u>: Claustrophobia, arachnophobia, fear of flying, acrophobia, agoraphobia, herpetophobia, hydrophobia, nycktaphobia, sidirothromophobia (my favorite), ergophobia, nosocomiophobia, etc.

<u>Physical Diseases</u>: Cancer, heart disease, multiple sclerosis (MS), cystic fibrosis, dyslexia, hemophilia, Alzheimer's disease, sickle cell anemia, etc.

Sex: homosexuality, bisexuality, transvestitism, exhibitionism, raping, pederasty, sadism, masochism, zoophily(bestiality), coprolagnia, sexual fetishes, sexual aberrations, etc.

Mental disorders: Schizophrenia, paranoia, psychosis, Tourette's Syndrome, insomnia, addiction (alcohol, nicotine, drugs, etc.), self-destructive, multiple personality, wife beating, child beating, hypochondriac, etc.

The Individual Personality Group

The positive ones: Good, happy, loving, honest, conscientious, reliable, brave, serious, stable, thoughtful, sensitive, careful, calm and collected, patient, sweet, congenial, humble, modest, courteous, agreeable, polite, gentle, gracious, pleasant, neat, organized, humorous, observant, philanthropic, generous, altruistic, noble, heroic, unselfish, loyal, smart, imaginative, creative disciplined, strong-willed, humble, modest, etc.

The negative ones: Hateful, evil, egotistical, jealous, lying, rude and calculating, greedy, conniving, scheming, self-absorbed, vain, miserly, obstinate, dastardly, conceited, know-it-all, resentful, suspicious, pompous, belligerent, callous, sly, cunning vicious, mean, mercenary, stupid, sullen, wily, violent, grappling, condescending, uncouth, blunt, messy, caustic, phlegmatic, pusillanimous, hotheaded, explosive, lazy, etc.

The neutral ones: Artistic (likes to paint, dance, sing, sculpt, etc.), sportsman, sports fan, spontaneous, attentive to detail, proud, conservative, superstitious, loves to travel, persistent, impulsive, talkative, sociable, shy, outspoken, deliberate, precise, proud, supercilious, submissive, effusive, enigmatic, mysterious, extrovert, dramatic, compromising, etc.

"These lists are very incomplete—they contain less than one per cent of the 100,000-140,000 genes in the human genome[6]. Some traits may appear in two or more groups—no attempt was made to create scientifically...

---

[6] A genome is a full set of chromosomes with its...resident genes.

publishable lists here." I added. "I made them so that we have something to use in laying out the theory. That is our only motive here—we are not writing a treatise on gene...taxonomy."

"What about your conspiracy theory" Estelle shouted from the other room. "How does that figure in with GSP?"

"All in good time, my dear," I replied. "We don't want to start on that till our genes have had some cognac and feel comfortable and slow to react to...shocking revelations.

Talents. A Cocktail of Genes

"Let's start by taking a look at the first category, the talent group," I continued. "I put them first because I thought we can dispense with them pretty rapidly—I didn't expect any serious arguments about them. After all, most of these talents appear to be hereditary, so genes have the first word there. Was I right? Do you all agree that these genes must exist? Do you see anything worth arguing about?"

"I got a little problem with the music ones, you listed there," Andrew said—he is an amateur jazz pianist. I think most of them are more than just one gene, probably a cocktail of genes, which must all be in high strengths, to achieve a specific talent."

"How do you mean?" I asked. "Can you give an example?"

"Well, let's take piano playing, about which I know something about. To be a good piano player you need good hearing to distinguish between sounds, pretty long and agile fingers, good left to right arm coordination, a flair for performing…"

"Even some athleticism," Paul interjects, "if you are going to jump on top of the piano, play with your feet, or chin or elbows, standing up, backwards—I've seen guys actually hit the keys with their rear end…

"That, too," I replied. "So the collection of genes for piano playing is increasing."

"You also need the proper personality for it," Christopher added. "Guys…waste their talent because they didn't have the persistence to practice enough—the…obsession to become the best. A lot of people can learn to play the piano, but the cocktail of genes needed to create a world class concert pianist, is large—and strength of 8 or better in all of them maybe necessary."

"I have pretty much the same beef with the gene for athletic talent," Paul said. "That must be a cocktail also. And, I think, a different cocktail—depending on the sport. Each sport needs special physical characteristics—all coming from genes if dad is right—and different personality, sometimes. I would even add the "killer gene"—they have been calling it the killer instinct up to now, but dad says it is a gene—as well. They say many people could play like Michael Jordan, but nobody had the killer instinct—sorry...gene—like he did. I guess using our current nomenclature, he had it with strength 10—the rest of them with much lower value."

I am sure you guessed it by now—Paul is the most...adroit in sports in the family.

"I think you got to grant this cocktail status to dancing—maybe some others, too," Christopher—our resident dancing expert—said. "It too requires good hearing, rhythm, special body traits like agility, quickness, jumping, etc., not to mention some talents (genes) for correlating music and feeling and, of course, the right personality, as well."

"Sampras just...

"Please Mom, don't say anything. We hope to watch the tape afterwards," Paul barely stopped Estelle from announcing the latest Wimbledon score—she was just coming in from the Den.

"All right," I said, "I'll agree with you guys that most of the talents listed are achieved not by a single gene, but by an amalgam of genes, all with a high strength value. As for the issue of the personality, I believe that comes in to make *successful* musicians, dancers athletes or whatever. I was talking only about the talent here, not the successful exploitation of the talent. Anyway, the additions are useful, they sort of introduce the issue of personality in this discussion—something we left for later."

"Objection!" Paul interjected, "I might agree that there are special "talent" genes—after all it is hard to see how you can become a singer, if you can't carry a tune, or a painter if you are...colorblind. But the issue of genes causing people's personality, that is a different ball of wax, and don't you try to sneak it in here, suggesting that we agreed on it. I object vehemently!"

"I object also," Estelle said.

"Me, too," Andrew added. Christopher just sat there frowning.

"Objections sustained!" I replied. "After all, I didn't bring in the issue of personality—Christopher did. I did let it develop a bit on purpose, and I am now glad that I did—it will prove helpful later in the development of the theory. Anyway, I now appreciate even more, how rare some of these talents are—they require many genes and high values for all of them—worse than hitting a…trifecta in the races. And, by the way, I also agree that the cocktail concept is applicable to most of the others in the talent group, as well."

"I still have a question on the talent genes you guys are discussing," Estelle said. "I hope you'll be kind enough to…humor me—after all I haven't asked too many questions up to now, and I see you guys are about ready to leave this issue."

"OK, go ahead," I said.

"Clear this up for me,' she said—I knew there was trouble when she faked…confusion—"but I always thought that the reason that the children of athletes or musicians turned out to be good athletes or musicians was the fact that their parents encouraged them, and helped them along. After all, successful athletes or piano players, know some of the…tricks of the profession—they can pass them on to their children. Then, children always try to imitate their parents—they start practicing from young. Now wouldn't all these factors lead to good athletes or musicians, and not some imaginary genes that you have come up with…after you retired?"

"You have a legitimate question, but there is no need for…pugnacity." I answered.

"Her question is actually a…regurgitation of the "heredity vs. environment" dilemma, the old nature vs. nurture routine," Andrew interjected.

"You can call it whatever you like, but that wouldn't be an answer to the question now, would it?" She replied.

"OK, OK!" I said. "Let us tackle Estelle's question by first looking at the extremes. Do you think it possible that Shawn down the street—zero athletic gene strength—could have been turned into an athlete, if he had been born in an athletic family—if somebody taught him the...tricks of the profession?"

"I guess not," she said...begrudgingly.

"Now, do you think Pele, the greatest soccer player ever, arguably the best athlete ever, would have lost his athleticism if he had been raised by a family of...accountants?"

"I guess not," she said. "But what if his parents were peasants from Bavaria and they stuffed him with so many...pork sausages during childhood that he weighed 300 lbs. at the age of 14? What sort of of...Paley, or whatever you called him, would you have then?"

"Even then," Andrew said, "he might have surprised you. You know, Maria Kallas—the opera singer—she was 300 lbs., I think, when she was discovered. That never stopped her from becoming one of the greatest ever. Of course, most opera singers are heavyweights so..."

"Actually," Paul interrupts, "I think Pele's father was a mailman and everyone knows they have to be pretty athletic to climb over fences to dodge dogs, and... "

"So," I said, ignoring all the interjections, "we agree that when the athletic genes are of strength 0 or 10, the environment has pretty near no effect on the outcome. The only question is whether athleticism can be improved, when the gene strength is in between these extremes. I think I'll yield the floor to our sports expert, Paul, on this one."

"I think the environment—rearing, mentoring, coaching or whatever else you want to call it—can make you a better player, a more effective athlete. It might convert you from a...loser to a winner, but it will never improve your athletic prowess," Paul said firmly.

"What about practice?" Estelle insisted. "That is part of the nurture thing also, isn't it? Everyone knows that you can get better with practice!"

"Sure, you get better with practice, mom" Paul replied, "but only up to the limits of your ability—up to the strength of you athletic gene, as dad might say. Continuous practice might help you reach this limit, but after that…ziltch—no improvement. I hate to say it—I'm no advocate of any GSP, myself—but dad is right on this one. I think we ought to move on—we are starting to beat on a dead horse now."

"Fine" I said, delighted to have Paul in my corner, even for just this one fleeting instant.

THE VIOLENT SURVIVAL SUB-GROUP AND THE CONSPIRACY

"We now move to the second group," I started "that of the survival/perpetuation group of genes. All genes are there for the survival and thriving of the species, of course. But this group is the most directly involved one—the most important one for the task. I split it up in two sub-groups there, the hard (or violent) one that deals mostly with basic…raw survival, and the soft subgroup—Love and Culture. The hard subgroup, labeled Life and Territory, does it by brute force, if necessary—the soft one by creating a strong, cohesive social unit."

"Violence is a big part of the first subgroup. Survival is top priority—all methods are allowable. Nice…genes finish last—to paraphrase that famous football coach's slogan. To put it bluntly, these genes are encouraging you—nay, they are…compelling you—to survive personally, and to help the species survive and flourish, *every which way you can*. Let me reiterate this once more. These are the most important and the strongest genes of all—they demand…satisfaction. The rewards are immense if you…obey—complete happiness and fulfillment in life. If you disobey, or fail to fulfill their demands, unhappiness and even…mental depression may be the order of the day. Suicide is not totally out of the question in extreme cases of inability or failure to fulfill their dictated tasks."

"Are we finally going to discuss this conspiracy theory of yours?" Estelle asked. "This is got to be the group of genes responsible for it, right?"

"Yes, we are going to discuss it." I replied. "As for which genes are responsible, I'll let you deduce it from the discussion. By the way, I hope you will find that it was worth the wait."

"I don't know about that," Estelle replied. "Don't forget that we have arguing about your ideas for years. I doubt that anything you present will be…astonishing to me, unless you put some new…spin on it" she added.

"Well, this time your objections will be written down and recorded for posterity," I said. "However, so will my rebuttal. So weigh your words— you don't want to come off mean-spirited in the book."

"The soft group are…nice genes, right?" asked Paul. "They wouldn't be involved in anything like a conspiracy, would they?"

"There is no such thing as nice genes or bad genes, just genes with a mission—survival and perpetuation of our species in this case," I replied. "The soft group are involved in encouraging you to do some…nice things, but only for the sake of survival, not because they have some code of…ethics. Falling in love may sound romantic and noble, for example— the genes want you to think in those terms—but it is there for the practical reason of mating and adding to the species. Culture is the adhesive that binds the tribe together. Language, the glorifying of heroes, the stories of…altruistic acts, the various rituals (birth, manhood, marriage, burial, etc.), all these are part of the culture that give the tribe a common purpose—a sense of togetherness."

"And to answer Paul's question," I continued, "the soft group is quite involved in the conspiracy, just as involved as the hard group, though the hard group's…means might be viewed as more detestable"

"I'll like to hear a definition of this conspiracy you guys are talking about," Paul said. "The only thing I have seen so far is a small reference to it in *The Racist Gene* story. What exactly is this conspiracy—by whom and against whom?"

"It is a conspiracy of our genes against…ourselves," I replied. "It is the control they exercise on us to perform various acts which maybe detrimental to our own existence, camouflaging it as grand and noble, even virtuous or meritorious. Worst part of this conspiracy is that nature has placed in us a gene (or genes) which renders us incapable of recognizing this conspiracy—voids our ability to reason it out.'

"And why should nature come up with such a conspiracy of ourselves against ourselves?" Paul asked. "I presume you agree with me that our genes are...ourselves—they work for us—they are...us, so to speak—they live in our DNA, right? Why should our genes want to do something detrimental to our—and therefore their—existence?"

"You got a lot of questions there," I said, "and some of the answers are yes, and some are no. Yes, our genes are us—they are housed in our DNA—but NO, they don't work for us, not full time, anyway. They work mainly for the SPECIES. And this is one of the aspects of the conspiracy. When there is a dilemma, when they must choose between us or the species, the genes will always opt for the species—much to the detriment to the individual, or even his tribe."

"So there are two aspects to the conspiracy?" Estelle asked. "This I have not heard before. What is the second aspect—if I may ask—or is it not...due time for that, yet?"

"Let us understand the first aspect first," I said, "and then we will get to the other aspects. There are more than two by the way, so you must be patient."

"OK," Andrew said, "I still don't totally understand how we can behave against our own self interest."

"Let me give you an example from animal behavior," I said. "I am told that the male praying mantis gets eaten up by the female after mating— nature chose this...deadly path to ensure their...survival. Yet, the male keeps mating—its genes compel it to—or the species becomes extinct. Its behavior is detrimental to its personal existence, but beneficial to the con-tinued existence of its species. It is not aware of its genes' conspiracy, any more than we are. We may be a more advanced species than these insects, but often our situation is the same—we must sacrifice ourselves for the good of the species. Lucky that the society is—unwittingly—part of the conspiracy, so such acts are considered noble, altruistic, heroic and the like. As such, they are rewarded accordingly, even though the person may be dead—no longer around to enjoy the rewards."

"OK, I understand it now," Andrew said. "You are speaking, of course, about the various altruistic acts when someone sacrifices himself for the common good; heroic acts where the person ignores astronomical odds against himself, and plunges into some act to save a group of others. Come to think of it, most of the guys who survived such acts, do not know why they…performed them. They don't appear to have…reasoned them out, they usually refer to some…force which seemed to push them to them— they seem surprised at…themselves for having done them. Is it because they did it under the influence of the genes—is that the conspiracy? "

"Exactly," I replied. "The survival genes dictated the action. Most of these genes are usually dormant—asleep, so to speak. But the minute some threat is perceived, they wake up with a vengeance. In an instance, the individual is placed in a state that I like to call…*genal*[7]—at the complete disposal of the genes. In this genal state, he loses his ability to reason, except in a skewed way, which serves the genes. He is not aware of this— and this is a key part of the conspiracy. In essence, we live two kinds of lives. One kind when the genes are dormant and we can think logically with our brains, and the other when the genes are in command, and our brains are incapable of rational thinking."

"Are there degrees of this genal state or do we all have it with the same intensity?" Andrew asked.

"There are degrees in all things human," I replied. "These genes have strengths, as we discussed earlier, and their effect has degrees. The genal state is there, but the fanaticism in…obeying the genes varies with each individual."

"This genal condition, as you call it, does it only happen to individuals?" Christopher asked.

"It can happen to a group of people, it can even happen to a whole tribe, or nation," I answered. "All it takes is a perceived threat—not even a

---

[7] Genal: The state of mind when an individual is under the spell of one or more…genes-his brain no longer capable of rational activity-though he is unaware of it.

real one. If the entire tribe goes ...genal, you have a tribe on the warpath. Having a country—or even countries—with most of its people in a genal state is not all that unusual, it happens every time there is a war. That is why it is so difficult to come to some rational agreement between the warring nations—rationality is on...vacation. A cooling off period with a cease-fire might help, but not always. Nations sometimes stay in a genal state for generations."

"OK, I fully understand this perpetual genal state you are describing—I work with Jake, the Palestinian guy," Paul said. "But returning to the idea of the gene conspiracy, I have some questions I would like to ask."

"Shoot."

"Say the danger is real and it is against yourself or your family. The genes wake up, you are now...genal—the genes dictate how to face the threat. Where is the conspiracy here? I don't know many people who would succeed in repelling danger with...logic. Personally, I'd rather go genal. So here's my question. Where is the conspiracy? Heck, I think the genes are helping you here—going genal may be your only chance, especially if you are a coward."

"You are absolutely right," I said laughing. "Going genal may be helpful in repelling the danger. It may make you physically stronger—even fearless. The conspiracy is still there—you are genal but unaware of it—but its consequences may be good, if the danger is repelled. It may also be bad—you die and it was the genes' work, not your own.

"What about in the case of threats to one's tribe or country? Couldn't going...genal be helpful then, as well?

"Yes, it could." I replied. "Look, I am not here to argue whether it is better to go genal or stay...ungenal, in situations of real or imaginary threats—you could make a case for both of them, depending on the threat. The point I am trying to convey is that you have no choice—you go genal, and hope for the best. And when you do, you are no longer master of your own fate—you march at the sound of the drums of the survival genes, like a zombie, much like the heroes of the short stories who had...defective genes. You must always remember that the genes work for humankind, not for you, not even for your tribe. This is one part of the conspiracy, though not the most dangerous part—that part will come in a minute. First I want to get your agreement on the existence of this genal state."

"OK, I will agree to it," Paul said, "just to keep this thing going." The others soured their faces and muttered some sounds of agreement under protest.

"I'll take that as a yes, from all of you," I continued, "and move right along."

"We now come to the second aspect of the conspiracy, the one that causes a lot of problems in our modern societies."

"Why don't you present these ideas with a typical gene from this group—it will make your presentation a bit less...pedantic," Christopher said.

"Actually, there is an introduction to these ideas in the last of the short episodes –the one dealing with racism," I said. "Why don't you guys read it at this juncture—I got to go off on a bathroom break for five minutes."

I found them still reading when I returned.

"OK," I said. "Are there any new questions? Did you find Estelle adequately playing the...devil's advocate in that short exchange?"

"She was fine—asked the right questions," Andrew said. "I got to admit that I still haven't really absorbed the hard realization that if this gene theory is correct, we are all born racists, and we are...normal. Of course, normal or not, it is still contemptible, so the task ahead is quite difficult for humanity—if people are to coexist peacefully."

"I find it very hard to accept," Christopher said, "that we are all born with some other pretty nasty tendencies as well—always assuming that Dad is right, of course."

"But also some really good ones, too." Paul remarked.

"You can't characterize them as good or bad," I replied. "The survival genes—like all genes—evolved over millions of years for one and only purpose, to maximize the chances for survival. Now some of them are presently thought off as...nasty, some may even be...embarrassing. But we got to face up to the fact that they are there, all the same. Woe to the person, who has them defective—sometimes, even very high value can cause serious problems.

"What kind of problems?" Andrew inquired. "Is this the second aspect of the conspiracy?"

"Yes, it is, so listen very carefully, I answered. "Let's reconsider the hard sub-group—the bunch under Life and Territory. Here's the problem."

"These genes evolved over billions of years, to aid in the survival of the species in primitive, tribal societies. As far as we know, there haven't been any changes in human genes over the last few thousand years—though societies have changed dramatically, especially in the last thousand years. So, you see, these genes are trying to protect us from dangers that may no longer exist—they are trying to maintain a status quo that is no longer there—no longer desirable. In some cases—the racist gene, for example— we actually wish that the gene would disappear. In other cases—gene of violence, say—we want it there, but only...part-time, active only under newly defined threats. This schizophrenic attitude creates problems in modern life, even when the genes are only strong—never mind...defective. This is the second aspect of the conspiracy."

"Can you be even more specific?" Andrew insisted.

"Well, the discussion of the racism gene—before the Symposium—is one example, of course. Let's continue with the gene of violence, as a second example. In war, the guy with a strong gene for violence is a candidate for a...medal. In peaceful times, he still tries to solve his problems using violence—he is a candidate for...prison. Having a strong violence gene may have been beneficial as late as the 1900's, but no more. Societies now demand that the rule of law reigns supreme—that people find non-violent means to solve their problems."

"How about an example of modern living where the violence gene comes in and messes up the situation," Paul suggests.

"There are plenty of them in everyday life, but I will take one that is rather intriguing," I replied. "In the old days, violence was also used for climbing the social and political hierarchy—the violence gene was work-ing hand in hand with these genes."

"Can you elaborate on that?" Estelle interrupted.

"Surely," I replied. "Every generation goes through this struggle, chal-lenging the existing order, sorting out everyone's place in society. In our early days this struggle was mostly violent—pretty much the same way as in the primate (ape) societies. But in modern societies, this struggle is delayed, and it is expected to take place with non-violent means. This is also part of this aspect of the conspiracy. Young people with very strong genes in the hierarchical struggle area, could find themselves in trouble in the new societies—they could become juvenile delinquents. They might have ended up as leaders in primitive societies. but in these times they might find themselves in correctional facilities and prisons, for the rest of their lives.

"Have you discovered the cause of juvenile delinquency?" Andrew intervenes with a dose of sarcasm. "I say, this gene theory of yours is a... panacea for all of society's ills."

"I don't know about a panacea, but, yes, I do have an explanation for juvenile delinquency, though I doubt that it will be applauded by the...experts—it is not very comforting" I replied.

"Let's hear it," Estelle said. "I might need it in that city council meeting for the proposed sports complex—some are saying we need it, to reduce youth crime."

"It is quite simple," I said. "All the young people are doing, is going through their own violent struggle, to decide the new hierarchy. Their genes don't know that society has changed—that the youth must now postpone this struggle in order to attend school, learn new trades, and the like. The genes dictate that it's time for the struggle, and the kids with strong such genes are...compelled to participate. As I said before, society has changed, but these genes haven't—it is the second aspect of the con-spiracy. And by the way, let us not forget that according to the GSP, these genes were put there for a good cause—for survival. The sorting out of the new hierarchy during each generation is extremely important for stability in the society and its institutions—even today. The timing of the struggle is pushed up, and the means have changed, but the struggle must take place nevertheless, or society becomes chaotic. And incidentally, even the youth who don't take part in the violent struggle—find other ways to satisfy their yearning for violence—consider the ones that do as...cool."

"Is the creation of gangs related to this problem?" Paul asked.

"Well, that is one way to re-enact the struggle for hierarchy—create a gang and fight it out among its members. A nearby gang comes quite handy as well—fighting with them helps in deciding the strongest in each gang—the top dog and his subordinates."

"Dad might be right," Christopher said. "The apes also fight it out for this...hierarchy. In fact this struggle is so important that it precedes mating—I read this in a research study. I guess the females wait on the sidelines till the...pecking order is decided, and then they have their own fight to get the top male. It is nature's way of putting the...best genes forward, in both man or beast."

"Is that why Henry Kissinger was quoted as saying that power is the best...aphrodisiac?" Christopher interjected.

"He ought to know," Paul volunteered. "He must have enjoyed an increase in appeal, after becoming the #1 ranking male at the State Department. He obviously couldn't... get much, before—had to...beat them off with a club, after."

"But if the girls are naturally attracted to the highest ranking male, the gene that makes them so must be important and...peculiar to the female genre," Andrew said emphatically. "I don't even see this listed anywhere in the list that you passed out. In fact, I am thinking maybe there are some genes that appear only in males and some only in females. The hierarchical fight gene in males, and the fight for the top guy gene in females, are very good examples. Am I right, Dad?"

"Well, let me first say that I know the study Christopher is referring to," I answered. "I think it dealt with Rhesus monkeys. They were brought by ship from India to Cuba, some thirty years ago. They fought with each other for weeks, to establish territories and a…pecking order, and then they turned to the business of…mating. I guess nature feels that procreating is better done after these other factors are settled—all for the best survival of the species, of course."

"Now to Andrew's questions." I added. "The list was never meant to be complete—I mentioned this at the start. As for the question of special genes for men or women, I really don't know. My recollection is that biologists claim that all genes are in both sexes, and they are activated accordingly."

"Some are definitely non-existent in me," Paul protested. "I don't feel the slightest attraction to any…high ranking male."

"Nor me," Andrew and Christopher shouted in unison.

"I got to confess, I felt quite proud when your Dad became…chair of his department," Estelle said. "I definitely felt more attracted to him the day it was announced. But, anyway, I still don't know how to vote on the sports complex. Does it help with youth crime or not?"

"From the point of view of the GSP theory, it seems to help," Christopher volunteered. "Dad and I have discussed this before. It offers a way to simulate the hierarchical struggle—the teams, the competition for top player, the actual games—all of that can keep their genes…fooled—make them think they are involved in the struggle. Go for it, mom, vote yes on the complex," he concluded.

"I agree," I added. "Sports is probably our best substitute for…fooling the genes of violence—for youths and adults alike. The adults also have the genes of violence and wars aren't as… plentiful as they used to be. Sports is a good replacement—beat the other team to a pulp, kill them—I'll have the…usual, as they say. What a feeling of exhilaration if your tribe (sorry, team) wins—what a feeling of depression if it loses. Some people can't even sleep at night after a loss—their genes think that their tribe lost a fight for survival, and its existence is in jeopardy. The British keep mouthing that it is all for the…love of sport, but none of them really believe it. If the genes are…well fooled—and you want them so, in a...fake struggle—a sports contest is a fight for survival. A struggle to the...end."

"I heard, that in neighborhoods with gangs, most of the guys don't participate in sports." Paul volunteered. "Why do you think that is? Is their violence gene so strong that it can not be...fooled to think that sport is the...real hierarchical struggle? Or is their gene defective?'

"Either one would do it," I said. "It is not easy to fool these genes—they prefer gangs to sports. The boy scouts with their teams, competitions and other types of survival games, fooled the...pacifists—the guys with weak violence genes. The tougher guys...mock the boy scouts, effectively announcing that their genes weren't fooled by...sissy substitutes—that they need the real McCoy."

"Now I see why dad said the issue is depressing," Andrew said. "If the theory is correct, the problem of crimes among the youth may not really have a solution. All these youth counseling programs with...role models, early...mentors, and the like, are headed down the road of only partial success, if not outright failure."

"Well, if he is right, we ought to start thinking about other ways to...pacify these genes." Christopher said. How about sending them to some island and let them go...wild on their own for a year or two—create artificial "Lord of the Flies" type environments? This ought to tell us, at least, who the strongest and smartest are in the bunch!"

"Yeah," Paul added. "At the end of the two years you can grant them...Diplomas in "Survival and Hierarchical Studies" and have the various Mafias (Italian, Russian, Korean, Chinese, etc.), interview them for employment."

Another option would be to send them to some boot camp—I heard they exist and are getting popular these days," Andrew said. "How about bringing back the draft?

"Guys, guys," I said. "We have gone way off the subject here, we are not meant to propose remedies for societal problems—only explanations of their origin. Let's take a brief break here, have some of Estelle's baklava, and then return. We got to turn our attention to the soft group now and I want to make sure that I can answer all your questions—these genes are the top-ranking genes in the gene…kingdom. Is there one last question on the hard group?

"I have one," Paul replied. "It deals with the genal condition—part of the conspiracy. How do the genes fulfill their mission if they force a country to a war that destroys it? Humankind has a lot fewer people when one country wipes out another, doesn't it?"

"That is a good question," I replied. "Sorry I didn't explain this issue a little better, earlier. As I already mentioned, the genes recognize only one master, the entire species. They will sacrifice a person, even a whole tribe or country for what they perceive as the…good of humankind. And the good they have in mind is not only survival and perpetuation, but also evolution. Their mission is not only to perpetuate the species, but also to improve it. If a strong tribe wipes out a weak one, they don't care—that is part of the game they are playing. I tell you, this conspiracy, this brutal devotion to evolution, causes a lot of grief—its goals may well be too anachronistic for our modern societies."

"I also have a question," Estelle said. "How do you explain death using this GSP theory of yours?"

"Death?"

"Yes, death." She replied. "It is something we all do, so there must be a gene that causes it, that's part of your theory, right? So how do you explain it? Why is this gene there? If these genes are so keen in having the species survive, why don't they *eliminate* this gene of…death? Wouldn't the species survive better if we all…lived forever?

She looked a bit cocky—she thought she had me… cornered.

""OK," I replied, "First let me admit that these are excellent questions. Thank you for asking them. They will help in illustrating the theory, and elaborating further on the gene conspiracy. Now to the answers."

"I stressed at the start of the development of the theory (and I re-iterated it again a minute ago), that the genes are not just keen in having the species survive, but also evolve and flourish. They pursue this task with vehemence—not hesitating to sacrifice even the individual who...houses them. We all die, as you wisely pointed out, and thus—invoking the GSP theory—death must be an integral part of their...approach to this goal. I am sure that the genes believe that the species evolves and flourishes much better <u>with</u>...death, than without it. I suppose that forcing the living to reproduce, age and die, serves the species better in the face of perpetually changing conditions over the millennia. Don't forget that mutations play a big part in the scheme of evolution, and that death and re-birth is necessary for their occurrence."

"So death is another aspect of the gene conspiracy?" Estelle wanted to know.

"I am afraid so," I replied. "Killing us in the name of evolution is something they do routinely—with no remorse. The way they see it, *they must keep destroying us, in order to save us*. That is the epitome of their conspiracy."

THE SOFT SURVIVAL SUB-GROUP

"Are you making the usual claims for this group as well" Paul asked when we returned. "Are all these activities described in your list dictated by genes?"

"Yes, and many others, not included in the list," I replied. "These genes are just as important for survival as the violent group. Some, in fact are so important, that if a threat is perceived to its dictated activity, the person is placed on the…genal condition—and all its consequences."

"What about being in love?" Christopher wanted to know. "Is that some kind of…genal condition, do you think?"

"I'd say that if falling in love doesn't meet the definition of a…genal condition, I don't know what does," Paul butted in. "I haven't known any-one in love who can see straight, let alone reason things out logically. It is a state of stupor, that's what it is—stronger than genal. I wonder why the gene—if it exists—makes you behave that way."

"It is the gene's way, to ensure the propagation of the species," I replied. "The gene locks you in with a partner, ensuring a mating and possible procreation. You got to do all this, without much thinking, in a zombie like state, until the children grow up a bit, anyway. Yet, you never stop thinking that you are in complete control—it is all part of the conspiracy."

"One thing I noticed," Andrew remarked "is that suicide is quite preva-lent in teens, and often due to…unrequited love. Could this have some-thing to do with the love type…genal condition?

"I think you got something there," I replied. "When sexual maturity is reached, the genes expect you to start…adding to the species—it is your reason for living. So you go ahead and fall in love, happy that you are about to fulfill your paramount task, and all of a sudden your partner betrays you. If your genes of procreation are very strong, they cause an intense sense of…failure and despair. You start to feel unfulfilled, useless—not fit to carry out your part in the tribe's perpetuation. Such feelings, can drive you to suicide. I hope you all see how the genes all conspire to make you their pawn. As I explained, they have to try to punish such failures—inability to add to the tribe's population is anathema to them. On the other hand, if all goes well, happiness will reign supreme. Not that you are any more in control—the gene conspiracy is still in effect—but at least you are happy and contended—not suicidal."

"How about expanding a bit on religion?" Estelle asked. "I know you don't have much of a…religiosity gene—as you call it—but you still respect its importance in the survival of the species, right?"

"My religiosity gene is probably…defective," I replied. "Even so, I am stating here categorically—and I have done so religiously over the years—that religion is a big part of culture—perhaps the biggest. The religiosity gene is very powerful—anyone thinking otherwise is a fool. A major reason for the demise of the communists was their obsession with eradicating religion—*the opium of the people*, as Lenin called it. They even thought that their political theory will satisfy the religiosity gene's demands. But as soon as communism fell, religion made a comeback—the gene had the last laugh."

"So you admit that the religiosity gene is important?" Estelle persisted.

"Yes, I do. In fact, I believe that Nature thinks so highly of this gene that it granted it a very high rank in the culture group. Don't forget that the two most revered people in a society are the political leader and the religious leader—that's how big this gene is. In some societies, this gene completely dominates culture and politics—it is the main culprit for the culmination of…theocracies."

"Well then, how come there are so many atheists around?" Estelle inquired.

"The gene of religiosity does not advocate a specific set of beliefs—it just dictates that you should embrace *some such set.*" I replied. "If your gene of…*logicality* (the gene responsible for your deductive prowess) is high, you may be unable to accept a belief in God and the rest of the package that organized religions…spoon-feed you. But if your gene of religiosity is also high, you still long for some set of beliefs, and a group to share it with—sometimes even with a little fanaticism. That is why these various non-religious groups (atheists, agnostics, infidels, free-thinkers, etc.) exist, and in some cases even flourish, especially when they are organized into a cohesive unit."

"And, what, if your gene of religiosity is…zero, as it is in your case?" Estelle inquired.

"I suppose you end up like me, totally indifferent to organized religions or non-religions." I replied. "People like me can not form or belong to such a group. Their weak—or defective—gene of religiosity renders that an…oxymoron. So you see, the opposite of high religiosity is not advocacy of God's non-existence, but complete indifference."

"What about all the crimes committed in the name of…God, or…religion?" Christopher interrupted, changing the focus of the issue. "What about the Crusades, the Inquisition, the religious struggles in Northern Ireland, Indonesia and other places?"

"Well, if a threat to the tribe's religion is perceived, the violent survival genes wake up with a vengeance—the person is ready to kill to defend the tribal beliefs. The genal condition overwhelms the tribe and atrocities are readily committed in the name of…God or religion. Such atrocities may amaze the uninitiated, but not the believer in the GSP. It is…survival gene business as usual. It may be that the gene of religiosity is…pious, but to the basic violent survival genes piety is as foreign as…niceness."

"It is still amazing, though," Christopher added. "I still can't believe how some people become so fanatic as to engage in massacres. Don't they see the contradiction of killing in the name of their God, or their religion?"

"Not while in the genal condition'" I replied. "That is all part of the conspiracy. The genes don't care about logic, dilemmas or contradictions—they are only interested in survival."

"How important is the totality of the culture genes for the survival of the species?" Christopher asked, returning to the central topic of conversation.

"Just as important as Life and Territory" I replied. "There is really no...tribe without it. If there were it would soon disintegrate—unless it developed a culture, and fast!"

"Dad is right on this one," Andrew said. "You heard of...cultural genocide—Native American have complained about it, often enough—it is a threat to their existence. Wars have broken up in many parts of the world, because of disputes concerning languages. Europe's main fear for the last 30 years or so—since the US started to dominate the world—has not been so much their possible subjugation by military or economic power, but by culture."

"This explains why France keeps passing laws to avoid this trend," Paul said. "I heard that recently they passed a law against shop signs in...English. This may also explain their...grumpiness against Americans—their refusal to speak English even when they are fluent in the language"

"They are reacting to a perceived threat to their survival—a threat to their culture!" I said. "Tribes have gone to war for such threats—not merely passed laws prohibiting...signs. The first step is to...ridicule the invading culture—the French have gone through that already—then comes the laws and finally...all out war."

There was some silence for a while—everyone appeared to be gathering his or her thoughts and trying to make some sense out of the theory.

"Are we about finished with the Survival genes?" I asked.

"No, not quite," Estelle said. "I would like to ask a few general questions if I may."

"You may," I said, though I knew that what was coming, was not going to be very pleasant.

"You see, I don't totally disagree with the overall idea that there maybe genes that encourage us to do these things," Estelle said. "But I find it hard to accept this tone of...inevitability—this lack of...choice you are implying—no matter how strong these genes are. Then, also, what about love, what about beauty, what about compassion in life? You make it sound as if we were…robots, automata—slaves to the dictates of these genes, even when we are not in the genal condition."

"Yeah, Dad, where is your…romanticism here?" Paul added. "Aren't you…trivializing the issue by implying that we fall in…love because some…gene said we should? And aren't you justifying our violent actions by also saying that a…gene made us do them? Aren't you implying we have no…feelings, and no choice in the matter?"

"Look here," I said rather irritated—I have a pretty strong temper gene, even though I neglected to put it on the list. "I am just calling it as I see it. There are plenty of books out there, full of…romanticism—if that's what you want to hear."

"What kind of books? Which ones are you…belittling now?" Estelle asked.

"Well, to begin with there are all these guides-to-happiness type books. They are usually full of meaningless statements like "live every moment to the fullest," "Live each day as if it is the first day—or is it the last day—of your life," "Look at things as if for the first time," "count your blessings," and all the rest. Only the authors of such…cliches can possibly become happy—if their book sells big."

"Then there are books by authors who are sprawled out on their death beds, spitting out profound wisdom like "most important thing in life is…love," or "I wish I spent more time with…my family—it's the only fulfilling endeavor in life," or "to ease your stress and unhappiness, go out and do…volunteer work—helping others, brings meaning to your life."

"What is wrong with such advice?" Estelle asked. "I always found it quite uplifting."

"There is nothing wrong with the advice—nothing…new either. What I am doing here, with the GSP, is address the deeper issues—the why's. We all know that love is important, but why is it? Why are we happy when we have children, when we help others, or do all these things that these authors are suggesting, while looking out at the horizon with glazed eyes—making contact with some…divine authority?"

""You got an answer for those things?" Christopher wondered. "I told you guys, Dad is on to something, here."

"Well, yes, I do," I said. "It is quite simple, but somewhat controversial"

"Let's hear it then," Estelle said. "Though, I think I already know the answer."

"The simple answer is that...genes want us to do these things—to aid in the survival and the successful perpetuation of our species—how many times do I have to say the same thing? If you do them—fall in love, repro-duce, raise children, help others and the culture—you will be re-warded with happiness and a sense of fulfillment. If not—you already know what happens—I just said it a few paragraphs ago. Now when these...romantics say it—in their books...oozing with hypocrisy—people reach emo-tional...orgasms. When I say the same thing but add an explanation of the "why" (the genes...expect us to do them), the...accusations start fly-ing. You are too...this, or too...that. You turn us into robots, or...automata. Well, does any one have a better answer for the "why's"? If so, let's hear it."

"You know what I think?" Andrew intervened, "I think Dad might be right. But I also think, that the genes don't want us to know these ...whys. They are fooling us into thinking that we made a conscious decision—after serious and...profound thinking. But why does the...profound thinking always ends up with the same conclusions, and we perform these same acts? That right there is an indication that we may be programmed to reach these conclusions—but at the same time think that we men-tally...worked them out. That is the conspiracy, right Dad?"

"Precisely, " I said, "though I am surprised that your genes are...letting approach these profound truths—it is most likely due to my insistence here. As an example of what Andrew and I are saying, let me ask everyone a simple but also abstruse question."

I stood up, took on the airs of a deep thinking philosopher and then uttered the phrase,

"Why do we find…beauty in a nicely shaped naked body? In fact, this may be one of the universals—pretty much everyone does, with few exceptions. Why???"

Complete silence. Everyone is squirming, rolling their eyes, and scratching their heads.

"OK, let me help you out a bit," I said. "Why don't you get the same sense of…beauty when you look at a naked…cow?"

"Because…because…it gives you pleasure to look at a human body—no pleasure from the cow," Estelle whispered, without much…conviction.

"That's not an answer, Mom," Paul said. "Dad wants to know why. Why does it give you pleasure?"

"Well, because it's beautiful," she replied laughing—realizing her answer was a…tautology."

"Let me ask you some other questions. Why do we find stories involving love beautiful? Why is there such immense pleasure in eating, in sex, in having children? Why do we love our children, our grandchildren, children in general? I won't go on, the list is endless."

"Eureka," Christopher shouted. "It is because…" here he started scratching his head, pretending to be thinking very hard, "because…the genes force us too. By George, I think I got it! I got it, I got it!," he was singing to the tune of the familiar song from <u>My Fair Lady</u>, dancing around the room in faked exhilaration."

"Amen," I said "Our notions of beauty or ugliness are totally…gene dictated—it follows from the GSP principle. It is no co-incidence that humans pretty much agree on what is beautiful and they are attracted to it, nor that they agree on what is ugly and they are repelled by it. The attraction and repulsion serve to plant the right genes to the next generation—it is all part of the gene blueprints for improving the species. And the sooner everyone takes one step towards accepting this, the better it will be for humanity. There may even be a few solutions to some of society's problems, if we accepted this basic truth and went from there," I concluded.

"And by the way," Christopher adds, "I don't think that Dad's answer here takes away any of the beauty of a naked...babe, makes falling in love less divine, or takes away the pleasures of sex and all the other...basic activities. The genes may have...dealt us all these cards without asking us, but I, for one, want to thank them—let them know of my appreciation. I am certainly glad I am not programmed to fall in love with a...cow. That would really be a...nasty conspiracy."

The others made some "oh, well," type noises and re-adjusted their seating postures.

"Any more questions?" I asked.

"Yes," Andrew replied. "What about our sense of morality? Is that dictated by a gene, also? I am guessing that your answer is yes, but I'd like you to elaborate."

"Thank you for bringing it up, and, yes, my answer is that our sense of right and wrong is totally dictated by our genes—totally designed to serve their purposes. Again, it is a simple application of the GSP principle. Most people agree on what is right and what is wrong—if it helps the species, it's right, if it hurts it, it's wrong. Some people put a mystical aura on such things—they argue that morality comes from some religion, or from a profound thinker who formulated a set of moral principles using his brain. The harsh truth is —I am afraid—that all of that is part of the gene conspiracy. The profound thinker would *always* come up with the same general moral principles, and these principles would never contradict the dictates of the genes. That is why many people insist that there is...nothing new under the sun; the words may change, the notions may be camouflaged as new, but the essence of morality remains the same—the dictates of the genes have not been altered."

"Are you claiming then that all societies have the same moral principles?" Andrew persisted.

"Yes, with some variations which help the species survive under local...peculiarities. For example, if the cow is of paramount importance for the survival of the tribe, it may be declared...sacred. If some animal carries a deadly disease, eating its flesh may be deemed...immoral. Such variations in the sense of right and wrong do not contradict the general principles of morality set out by the genes. In some cases the variations may appear cruel, but morality dictated by genes is not inherently...nice—just efficient. And, as always, if the tribe is placed under a...genal condition, from some perceived internal or external threat, morality can change dramatically—cruelty is easily tolerated, sometimes even elevated to the status of...good."

"Do you have examples of that?" Andrew wondered.

"Plenty, but I don't want to dwell on the issue." I replied. "Let me just give you one. Putting American citizens of Japanese origin in concentration camps during World War II was an immoral and most likely illegal act. Yet it was accepted as... right, and even approved by the US Supreme Court as...legal. Now that we are off the genal condition, we consider this act as...evil."

"Now are there any more questions?" I asked again.

"Just a little one, and then we can move on," Paul said. "I am sitting here wondering about this notion of the gene conspiracy. That the genes are in control of our lives and we haven't the foggiest idea that we are not—in fact we think that...we are. That the reason for that—you claim—is this gene, which prevents us from being able to recognize this conspiracy. Which, of course, makes this gene, a key part of the conspiracy. Am I right so far?"

"You betcha," I replied.

"So here's my question. How come this gene...permitted you to understand and expose this conspiracy?"

"Very good question," I responded. "Actually, I am not sure of the answer. The only explanation that I can come up with is that my gene in question—call it the cover-up gene—is...defective. Not totally defective, partially, I think, for it has taken me many painful years to do it. In fact, Christopher here, and possibly even Andrew may have inherited this defect in the gene—considering their behavior and what they have said today."

"So, because I am arguing with you, you are accusing me of possessing the cover up gene, healthy and with high value. Is that right?"

"Yes," I replied. "But why take it as an...accusation? You—and Estelle—are more normal that the rest of us here. The pity should be on us—it is tough living with this...burden!"

"Maybe it should be, but somehow I feel shortchanged," he replied.

"Well, cheer up," I said. "We are about to take a brief break, just to gather our thoughts for the next challenge at hand. I think it is a good idea to peruse the survival genes one more time before we move on to the next group. I challenge you to ask some tough questions—even try to...ridicule this theory if you can."

Fun with
Survival...Cursing...Gossiping...
Celebrities

"Boy, I sure wouldn't like to have that "gathering" gene." Andrew started, after the pause. Is that the one that turns you into a...Bag Lady?"

"That's the one that causes you to want to shop," I replied. "It is mainly a...woman's gene. I think. Nature put it in them to encourage them to go around the tribal lands, gathering berries, wild fruits, whatever. Men did some of that as they traveled around for the hunt, but they may have been outside their territory in which case they were...stealing. So the gene for stealing is there as well—one does not die of starvation because it is...wrong to steal."

"Now wait a minute," Estelle says, "Don't give me that woman's gene rubbish, you love to shop when we are on...vacation, and don't you deny it."

"That means that men did their gathering on the way back from the hunt, when they were on...vacation," Christopher said.

"So, I was right about the Bag Lady," Andrew asked.

"Sort of," I said. "If you have this gene with value 10, or if it is defective, you could be an...obsessive shopper."

"If that's the gene you are talking about," Paul said, "They have already found it—they even found a pill to...arrest it, when defective. It was over the Internet, not too long ago."

"What if the stealing gene is of very high value or defective? Do you become some sort of...Robin Hood, or the head of the Mafia?" Andrew wanted to know.

"Please, please, let me play in this game," Estelle says. "You turn into a...kleptomaniac. Right?"

"I think so," I said. "And if you gave all your loot to the poor then you turn into Robin Hood."

"And if you have a defective procreating gene you become a...sex maniac, a Casanova, right? With a defective hunting gene, you turn into a...bounty hunter." Estelle was going wild and...loving it. "With a bad stockpiling gene, a...pack rat. Bad fishing gene, you hang around even though the only thing you are hooking are old...boots. With a bad ritualism gene, you are always attending funerals, or constantly mixing up some witch's brew."

"Let me have a crack at it," Christopher interrupts her. "With a bad politicking gene, you become a Pat Buchanan or Ross Perot or a Harrold Stassen—a perpetual candidate. With a good love gene you turn into... Roberto Benigni—"I want to make love to everybody". With a bad singing gene...Tiny Tim. With a bad racist gene, you have a choice of a...skinhead or...Archie Bunker. With strong religiosity gene...Jim Bakker, the evangelist, or the... Pope. A strong machinating gene turns you into...Macchiavelli. With a strong interest-in-politics gene, you turn into...Sam Donalson, and if it's defective, you become the whole MacLaughlin...group."

"My turn," said Paul. "Even though I am still skeptical about this gene theory, I can certainly...play this game also. I'll go with low values on the genes—defective in the...negative way. Here goes; bad religiosity gene— you turn into...Dad here. With a bad procreating gene, you better get yourself some...Viagra. With a lousy machinating gene, you turn into a...Ralph Cramden."

"I think we played this game to its limit," I said. "Besides we have to leave some things for the reader's imagination. Are we ready to move on to the next group?"

"No, we are not," Andrew replied. "I see a "cursing" gene in this group, and I am wondering why it is part of the survival group—if there really is such a gene."

"And I see a gene for gossiping and one for interest in…celebrities," Estelle said. What on earth are they doing in the…survival group? You are not implying that you can survive by…gossiping your way out of a jam?"

"Or impressing your enemies with your knowledge of…celebrities," adds Andrew.

"Quite hilarious," I replied. "Let's just take them one at time. First the cursing gene. I am surprised that it is being questioned—I thought it obvious that it is a gene, after the late 60's episode on…coprolalia. Even primates use…cursing (shouting, looking mean and scary, etc.) to defend themselves—to…ward off possible attackers. People use it when they sense an imminent attack—it is a first attempt at putting your opponent in the defensive. It is also…universal—all peoples have cursed and cussed, and continue to do so—and that right there qualifies it as a gene for…survival. In fact the first thing people want to know when learning a foreign language is how to…swear in it. Their cursing gene…demands it—it feels ineffective in a foreign country. The other genes use…universal language."

"You talked me into it," Andrew said. "I've used it myself on occasion to scare people. I suppose if it goes bad, that's when you end up with Tourette's syndrome, right?

"Yes, or some other related disorder," I replied. The gene is usually dormant—it wakes up when danger is approaching. One possible defect is that its mechanism goes awry and it starts turning on at…random times. That's when you get these various…syndromes. It is fortunate that such disorders have been now properly diagnosed. In the old days, people believed that the…devil, or some evil spirit had entered your body—and was cussing through you. The church usually recommended exorcism in such cases—a rather…nasty experience."

"What about the guy who's every other word is the F word?" Christopher asked. "Does he have a defective gene?"

"I doubt it," I replied. "My guess is that he is doing it on purpose, to show his…toughness. If I am right and cussing comes from a gene akin to violence—verbal violence, of course—using macho type language announces your...readiness to attack—provoked or not. This actually becomes fashionable and highly imitated. It is most popular with people who make their living in the underworld—a form of...advertising your abilities."

"I got to use more of it myself in the basketball games," Paul commented. "That industrial league we are playing in right now is...torture—you can't drive to the basket without being hacked. I think I'll start swearing...before I drive—it might scare them off. Cussing them afterwards doesn't seem to help."

"It seems to work pretty well with...animals," Estelle added, "they seem to recognize as a possible ...threat. Just the other day, I scared off an armadillo in the yard, by shouting at it. He was being aggressive—showing his teeth at me and growling."

"You obviously...outcussed it, mom," Andrew suggested

"Now I'd like to discuss gossiping—if you don't mind," she resumed. "You men accuse us, women, of gossiping all the time, but when you guys do it, it's... exchanging information."

"Here I have some good news for you," I said. "Gossiping is another universal trait and when that is the case, the gene's origin involves...survival. That is the basic principle of the GSP theory."

"I thought so," Estelle said. "I know a lot of people who simply could not...survive without perpetual gossiping."

"It is not that, mom," Christopher said. "I think dad means that gossiping helps in the survival of the species, and thus it is a...necessary gene—we all have it. Myself, with strength zero or at most one," he added.

"Me too, me too," the other two shouted.

"There you are, still thinking of gossiping in derogatory terms—the gene theory is not yet...sinking in you," I said. "in the old days, the gene was extremely important for man's survival. It was the tribe's...mass media. People had to know of impending dangers, who is sick, who has some mental problem, who is unfaithful—all these could spell danger or instability for the tribe. Gossiping or information exchanging was very useful in matchmaking—also part of survival. Of course, women did most of it—the men were out on the hunt. But they too brought back news (gossip) of neighboring tribes. I think the persons with high gossiping gene value (the busy-bodies of today) were probably...revered in primitive times—they were the tribe's...local library. That is why everyone loves...a library," I added in jest. "That's where all the...gossip is."

"That makes plenty of sense," Estelle says. "I can see, though, that the development of the mass media has made gossiping obsolete. I suppose it is still useful in countries where matchmaking is still practiced—you can't get the information you want for that from the ...newspapers, you know."

"But how did it become derogatory in our time, if it was so...noble, at the start?" Paul wanted to know, thinking I had no answer for it.

"I think the...busy-bodies did it," I said looking serious, as if I was about to spout a profound remark. "It is not easy going from revered to...useless—the gene demanded...action. So they started...making up false information—quite colorful, of course, to regain the attention. It was all...downhill after that. The gene is still there, the gossiping must still go on, but its...image has gone down the gutter. That is why we don't feel proud about doing it" I concluded.

"OK, how about the gene that makes you interested in celebrities?" Estelle asked. "Aren't you getting somewhat absurd with this kind of gene specialization? Is there a gene for interest in some specific kind of celebrity? How about a gene for interest in notorious gangsters? Or a gene for fascination with female actresses who faked pregnancies?" She was on a...mocking trip now.

"There could be," I replied. "It would certainly explain people's interest in talk shows with such... titles."

"But let me get serious about this for a minute," I continued. "Interest in celebrities being universal, must also come from a gene connected to...survival—most likely in the cultural group. I am guessing that the gene evolved to make us interested in the lives of our heroes and leaders—political and cultural—the original tribal...celebrities. You had to know if the leader is sick—a struggle may be on for the top job. Maybe his wife died—big news for the available maidens. Another reason for following their lives was for...emulation—they were...role models as we now like to call them. The original celebrities were an integral part of the history and thus the culture of the tribe."

"Yes, but nowadays, nobody gives a damn about our political leaders—they are...material for stand up comedians." Andrew interjects. "The real celebrities are in the movies, fashion, sports—not in politics. Do you have an explanation for that? Has the gene changed its...interests?"

"It serves them right—especially the...buffoons." Andrew interjected. "Why, you know what Dan Quale said the other day..."

"The only explanation I have is that political leaders are now...ephemeral, and heroes...quite scarce" I answered. "The gene...longs for celebrities that last a while—become historical figures. Even athletes have longer celebrity lives than politicians these days. The gene never figured on...democracy. Democracy is a product of...reasoning, not genes. That is why it is such a difficult regime to implement, especially if the tribe is threatened and thus placed in a...genal condition."

"Is that right?" Andrew seemed...amazed. "I've heard of a lot of reasons why democracy is hard to implement, but that the genes...dislike it—that's all news to me. What do you think they prefer? Plain monarchy, constitutional monarchy, iron-fisted dictatorship, or maybe...anarchy?"

"Certainly, not anarchy," I replied. "That's why anarchical movements have failed. They are driven mainly by young people, who live in...tight hierarchical structures. They want a...free-for-all—a chance to become the...silverback[8] of the tribe."

"So what do the genes prefer, if they are not...enamoured with democracy?" Andrew insisted.

---

[8] Silverback: The chief of a gorilla group—so called because the hair on his back is usually gray/silver.

"I think we are off the subject," I replied. "Let's leave it for the end—the Rarely Asked Questions. Let's get back to our main topic—the interest in celebrities."

"Glad you are back," Paul remarked. "I want to know how you explain the immense interest in notorious…gangsters and criminals. Some of the most successful movies and books dealt with such people. Don't tell me the gene considers them examples for…emulation—role models for the young?"

"You got me there," I replied. "Maybe the gene wants us to have a keen interest also in what to…avoid—what…not to emulate—the positive, as well as the negative. Or maybe we like the gangster movies because the gangster's society is reminiscent of our tribal days—the type of society all these genes…evolved for. It might be a…nostalgia trip for them" I laughed.

"I've had it, I am going home, I am exhausted." Christopher said. "I was up all night with that patent search—just barely keeping my eyelids open. How about meeting nest week, same time? I wouldn't mind another one of Mom's dinners again and it would give everyone an opportunity to absorb all of Dad's new…ravings—can't think of a better word for it right now. I'm just kidding, dad, you know I'm pretty much with you…"

"OK," I agreed, and they all seemed relieved. "We are finished with this group anyway. The present meeting is adjourned. Rendezvous for next Sunday, same time same place. Be here!"

In case the reader didn't know, Sampras beat Agassi handily—tied the record held by Emerson for several decades.

THE PHOBIAS. TRISKEDEKAPHOBIA?

This next group is the …funnest of all," Estelle started us on, right after the bellies were satisfied, and we moved to the den for the dessert—melo-macarona—and coffee. "It has the…cutest names, especially the phobias group. I just hope I don't have any of these disorders—you always start wondering, when you hear the…symptoms."

"We are not going to discuss symptoms here" I informed her, "This is not a medical…seminar. We will simply discuss whether a gene could be responsible for their existence, and whether that can shed some new light on the issue."

"Assuming that we are starting with the phobias group," Andrew said. "I think most of us will agree that defective genes are responsible for most of these maladies. But why is there a gene related to say arachnophobia to begin with? Why did nature put such a gene in us, which, having gone defective—I presume—resulted in this pathological disorder?"

"I think that I know the answer—dad's answer—to that, but one that I totally agree with, myself" A rejuvenated Christopher jumped in. "Nature put in some fear for bugs in us—a normal fear, not an obsessive one—to protect us from their sometimes poisonous or disease-carrying bites. When this gene becomes defective—there comes arachnophobia."

"Thank you, that's my answer," I said. "We don't know, of course why these genes go defective. Sometimes you inherit them that way, other times physical environmental factors in the womb—or outside of it—cause it. We just don't know. We let medical researchers, and the nature vs. nurture guys argue this out, at their own convenience."

"I think that the key question here is whether psychological factors can cause it." Paul comes in. "Well dad, what sayeth you?"

"We'll get to that question later—I have already promised" I said, "it is very important and it applies to all the genes, so postponing it might help us discuss it more thoroughly. Are there any other comments about the phobias group? You all realize, of course, that the list is quite...partial."

"I've got a question," Estelle says. "What sort of phobia is this sidiro-dromophobia thing, anyway?" she asks.

"Fear of trains," Andrew replies, "it is from the Greek word sidirothro-mos—train."

"OK," Estelle says. "Now you said that all the phobias come from a gene which was put there to protect us from something—to aid in the survival. And this gene went defective, Right?"

"Right," I said. "Acrophobia, from a gene to make us afraid of heights, so we don't go next to cliffs and fall off. Fear of flying from a gene which prevents us from flying off like...Ikarus. Agoraphobia, from a "social" gene gone defective—the...opposite way—causing you to abhor crowds rather than joining them. And on it goes," I said.

"Fine," Estelle answers. "Now what is this "fear of trains" gene doing in us, anyway? There weren't any...trains during tribal times, were they? Or did nature...know that they were coming and put the gene there to warn us to be careful crossing the...tracks?" She laughed—she loved squeezing you in a corner like that.

"Any ideas?" I pleaded for help.

"Well, I think it is from a defective gene that deals with our fear of ser-pents—akin to herpetophobia," Andrew volunteered. "You see the train tracks (or the trains themselves) look like a...snake and the defective gene thinks that they are." He was having a hearty laugh while saying it.

"Ok, I got a possible one," I said. "I think this phobia does not really...exist. I read it in a list of phobias, but I saw the author was some psychologist with a Greek name. I think he put it there for the same rea-son that I did—the word has an intriguing ... musical ring to it."

"I'll accept that," Estelle answered, and I breathed a sigh of relief. "Who could possibly be afraid of trains? Everyone loves them—especially children.

You should tell that Greek psychologist of yours that it is more likely that the disorder is sidirothromomania, and not sidirothromophobia."

"If there is such a disorder as sidirothromomania," Paul added, "then I have an explanation for the phobia, which you were quick to dismiss. It is mom's gene of the mania gone...berserk—turning love into...fear—the phobia."

"You guys are expanding on this theory faster that I can even...absorb," I replied.

"You bet," Paul said. "And by the way, what happened to triskedekaphobia? I noticed you...sneakily omitted it, even though its name sounds intriguing, not to mention musical. I am guessing you couldn't explain it, so you decided to omit it. How can there be a gene that makes you afraid of the number 13? Many tribes didn't have a numbering system then, and even today some tribes do not use the Arabic numbering system."

"Yeah," Christopher said, "and I know people that are scared of breaking a...mirror, and mirrors are a relatively new discovery."

"And what," Paul re-enters, "What if a guy is scared of the number 13, and then on the 13th of November he wins the Lotto and becomes a multimillionaire, and after that he loves the number 13? Did his "hate 13" gene convert to a "love 13" gene? I think the theory is breaking down—it is turning into the...theater of the absurd. Too bad—I had actually started to like it—it was leading to some rather hilarious interpretations of human behavior."

"Put the brakes on," I replied. "This type of...brainstorming is leading you guys into the realm of the...absurd—not the theory."

"The pertinent gene here," I explained, "is the gene of...superstition, which I put it in the personality group—who would guess it would create such...havoc. We all have it to some degree—value 3 or 4 in most. The ones with very high value may, I guess, have some phobias—or the opposite—obsessions with lucky omens. These do depend on the culture and everyday life—they might even change from time to time. Not the gene— mind you—the...object of superstition. Coaches wear the same sweater when they are winning, others come out of bed just so, eat certain foods—the list can go on and on. The confusion comes from the fact that some phobias (acrophobia, agoraphobia, etc.) can be attributed to a specific gene of their own, while others (all the superstitions, for example) are manifestations of the same gene.

"I still don't get it," Paul replied. "Could you come up with another example to make it clearer?"

"OK," I said. "A guy who likes to rape young girls has the gene of…pederasty (the trait), and not a gene for raping a young girl named Emily. A guy who lies habitually has a strong gene for lying (the trait), and not a gene for lying about his income, or his position. In the triskedekaphobia case the trait is superstition; the fear of the number is the detail. Is that clearer now? Can we move on to the next issue?"

"Not quite," Andrew enters the picture. "And why, pray, do we have this gene of superstition to begin with? You just said we all have it. You have also being saying all along that in such a case of…universal gene possession, the gene must have been placed to aid in the…survival. Well, what danger does this gene…protect us from? "

"That is the toughest question yet," I replied. "But even if I can't answer it, it does not mean that an answer does not exist. After all, these genes evolved over billions of years—and my gene theory is at its infancy. Their existence may be due to some…extinct danger that I am unable to identify, or even imagine at the present time."

"Can I take a crack at it?" Paul asked. "After all, I started this whole issue—maybe I can finish it."

"Go for it," I agreed.

"It might be a type of…warning gene, warning of possible danger. Early man (oops, or woman) lived in an unknown environment. If he touched a…plant, for example, and then developed a skin disease, the gene caused him to stay away from this plant, and tell others of its…bad luck. It is a…better-safe-than-sorry gene, it helps in the survival game—Dad's theory is OK."

"Gee, that's as good an explanation as I can ever hope to produce," I said. "I am really impressed because it came from the hardest opponent of the theory—the original doubting Thomas. I will give you full credit in the...book. Dare I hope that you are now a convert, or at least ...softening?"

"It is also interesting to speculate on how some of these specific phobias originated over the eons," Paul continued without answering me. " Now I don't know how the number 13, or a black cat, became bad luck," "but the broken mirror is easy to guess. When mirrors first came out, people could not understand the...*physics* behind them—they confused their image on the mirror with...themselves. Breaking a mirror—they probably believed—would somehow ...harm them."

"He is totally right about the mirror," Christopher said. "This explanation jives with a Western I saw once, where a...savage sees his face for the first time in a ...brook. A guy threw a stone in the brook, disturbed his image, and the savage... went ballistic. Apparently he saw his image becoming distorted, and thought he'd been...done in."

"And a black cat is hard to see at night," I said—you can easily step on it and then you might just discover why it is...bad luck. As for the number 13, the New Testament may be the reason—was Judas the 13th apostle? But I am leaving this issue, I am moving on" I concluded.

"I got one more phobia, which, I am sure exists," Estelle says, "most men have it, and Aris here...purposely forgot it. Fear of commitment. Is there a gene for that as well?"

"Good one." Christopher said. "Dad?"

"I don't know why you guys accuse me of intentionally omitting phobias, I wouldn't do that—I love them," I replied. "Especially one that deals with...commitment—haven't I already committed over 40 years to Estelle here? Anyway, fear of commitment is not, I think, a phobia. People who have it, do not go berserk, panic, or lose consciousness from some... impending commitment. I choose to think of it more as a personality trait. That's where I put it, anyway, under the interest in commitments label."

"Even if it wasn't a phobia, it is sure to become one now—after mom suggested it," Andrew remarked, jokingly. "Many a man would love to…fake it—plenty of unpleasant relationships around. I could have used it myself in this…"

"Let's get on to the sexual group" Christopher suggested. "It might turn out to be the most hilarious, not to mention controversial."

"I am ready," I said, "unless someone has a remark on the various physical diseases like MS, Cystic Fibrosis, Hemophilia, addiction, etc. We do all agree that some gene has gone defective here—some have already been found, anyway. Remember also that we are ignoring the reasons or the "when" (before or after birth) for these defects, leaving those to medical science, and the nature vs. nurture…polemicists."

"I have no questions on physical disorders," Estelle said, "but I am still…intrigued by that nosocomiophobia you listed in the phobias group. I know that *nosocomio* means Hospital in Greek, so I am guessing that the phobia causes…fear of hospitals. Does it really exist, or did you make it up to impress the readers?"

"A little bit of both," I said. "If it doesn't exist, it should—I believe I have it myself. I simply hate to go to hospitals, after reading all about the various germs, viruses and bacteria…lurking in every room or hallway—ready to attack everyone in sight. Then, you got to admit that the word is a thing of beauty—a definite…adornment for any good phobia…thesaurus."

"Is that so?" she says…sarcastically. "And how did the gene for this phobia evolve? I don't recall reading that primitive tribal societies had…hospitals. It doesn't seem like a superstition either."

"OK, that does it," I said fuming with anger. "It is now obvious that you are not participating in this discussion in… good faith—you are not here as a devil's advocate, but as the…devil incarnate. This sweet-talking me about the phobia, getting me to confess that I have it—thus proving that it exists—and then hitting hard with the last question, shows that your intent is…evil—you are trying to…wreck my theory. Do you know what you just did? It is called...entrapment—totally unacceptable in any court of justice."

"Yes, Mom, that was awfully sneaky" Christopher said. "All that innocence about the...intriguing phobias and stuff, and then the knock out punch at the end, that is too much—I can see why Dad is upset."

"OK, I am sorry," Estelle said, "maybe I overdid it a bit. But the question remains quite legitimate. Your court of justice is the...reading audience out there. Your theory must be acceptable to their common sense—legalistic type excuses are not going to swing them your way, you know."

"I don't care," I said. "Even though I have an answer, I am not going to give it under this type of questioning—I am going to...take the fifth. Besides, I never said that the GSP theory explains 100% of all human...folly."

"That is quite obvious," she said. "Another question that may be unanswerable by your theory is why people refuse to answer questions when they are offended—your situation right this minute."

"That is not so hard," I said. "They have a gene that prohibits them from doing so when they sense...hostility."

"Let's take a five minute break to calm things down a bit," I said. "We need cool heads to proceed with the next task on hand." I added pompously.

SEX AND...GENES

"On to the sexual group, then, perhaps the most controversial of them all, as Christopher correctly noted," I started, after everyone returned to the Den. "We have been hearing...Freudian explanations for these for more than 50 years—psycho-babble as some call it. You might have noted that I put in several short stories dealing with some of these deviations in human behavior—they cause a lot of grief in day-to-day living."

"Let me use the same technique in dealing with them," I continued, "starting with the easy ones and moving on to the tougher ones. First, the ones we can call...disorders, without fear of being ..politically incorrect. I am talking about pederasty (or pedophily), zoophily or bestiality, sadism, coprolagnia and other forms of abnormal sexual behavior. Is there anyone here that feels that these do not come from some defective gene?"

"If we start with pederasty," Andrew said, "I'll have to go along with Dad. Especially after that Texas pederast asked to be...eunuchized—have his testicles removed. He claimed he was unable to stop himself—essentially admitting that he was under the influence of an uncontrollable force. That's what dad is calling a gene—a defective one in this case."

"Pedophily is most...loathsome," Estelle said. "A crime against the children of a society is the worst possible crime. Maybe we should re-activate...Devil's island and sent all the pedophiles there to live with...themselves. Do you think there is a special gene that turns you into a pedophile?" She asked turning to me.

"My guess is that there is no special gene for...pedophily—just some other gene gone...faulty. Maybe it is the procreating gene which got mis-wired somehow with the gene that makes us love children—the future of the species. No matter how it developed though, the disorder is very strong, there is no present cure for it—all methods of...analysis have failed. Someday there may be a pill that changes the pederast's brain chemistry, but until then, society will keep incarcerating them, or converting them into eunuchs—if they request it, of course.

"Pederasty makes my...skin turn," Christopher said, "I'm glad you followed it with zoophily, which is...outrageous—even funny. I actually knew a guy once, from Italy, who claimed he was happy...boinking a goat"

"Heck, according to the...Kinsey report, written some years back," Andrew added, "a good percentage of American women procure their sexual pleasure from male...dogs. I have seen lots of women walking around with dogs, but I haven't noticed whether they were...male or not."

"I got to say—now that oral sex has become a household word—that dogs with their big, muscular and malleable tongues, could provide a very decent cunnilingus for them," Christopher added.

"If you continue with this embarrassing subject, I will refuse to go on with the discussion—I might even...boycott this meeting," Estelle said.

"There is no need to go on," I intervened. The main issue here is that these sexual aberrations (pederasty, zoophily, sadism, etc), are caused by some malfunctioning gene or genes. We can hypothesize that some genes got mis-wired with some other genes (erotic genes with animal loving genes, for example, for zoophily), but how it occurred is immaterial. What matters most is that we recognize it as a gene problem, and we don't go off preaching about...sinful ways, Sodoma and Gomora and the like. Society has to protect itself if such disorders cause its members some harm, but leave things well enough along, if they don't."

"Exactly," Christopher added. "It's nobody's business if someone has sex with a goat—it may be by...mutual consent. Unless the goat is...underage—a federal offense."

"Let me move on at this point and ask a key question here." Paul said. "I noticed that so far, you have very carefully avoided discussing the issue of homosexuality, bisexuality, etc., concentrating mainly on what we all agree are...abnormalities. Well, the time has come to attack the...holy cow—the king of controversy—homosexuality. Would you even call it an abnormality? Would you attribute it to a…defective gene?"

"Yeah, Dad," Andrew said. "Let's see your…guts here—the issue is probably the…zenith of political correctness. I don't think you can postpone it anymore—you got to come out of the…closet and lay out your opinion, right this minute." He seemed pleased with his word pun on…coming out of the closet on homosexuality.

"I never avoided it," I said, "I put in a short episode—A Gay Life Style—at the start of this essay, showing my…serious reservations on the conventional wisdom of this issue."

"So I will come right out and say it," I continued. "I think all forms of homosexuality are caused by one or more defective genes. There you have it—loud and clear. I could be wrong—most people are when they try to explain it—but that is my explanation. The explanation fits well with what we know about it. And it is not judgmental—its practice among adults is nobody's business but their own, anyway."

"Why is it that I had an inkling that this would be your answer?" Estelle asked.

"Maybe because, that's been his answer on everything so far," Paul suggested. "Heck, he actually sounds like a…broken record. Actually two broken records—one that says a normal gene did it, the other, a defective one did it. "

"Well, yes, but I still got to make a case for it—you haven't stopped arguing. Otherwise we wouldn't have a…book here, wouldn't we? And by the way, you can add the phrase <u>Nature put in that gene to enhance survival</u> to your…broken record phrase list"

"Explaining sexual attraction is not as easy as some of the others," I went on, we saw that at the start of the Symposium. Christopher suggested that it could be explained by a…normal gene with various…strengths—Andrew made a case for two genes with…defective turn-on switches. I like Andrew's explanation, because scientists claim that we all have pretty much the same genes."

"It would also explain why some people are born as...hermaphrodites—possessing both sexual organs. The genes with defective turn on switches were the ones with the blueprints for making the actual sexual organs."

"Can they have sex with...themselves?" Paul asked.

"Incidentally," I went on, ignoring Paul comment, "scientists have already found some gene or genes related to homosexuality, but the whole issue remains somewhat nebulous. I vote for the...defective gene explanation, for an additional reason. The percentage of homosexuals appears to be much less than 50%, so the property is not in most people—not directly involved in survival. In addition, I can't find a convincing explanation for having a normal gene that leads to homosexuality in terms of...survival. Don't forget our theory here remains the GSP. But the fact that I haven't found one, doesn't mean it doesn't exist. So I lay out my conclusions here with some reservations."

"I got one," Andrew stood up. "It might be that Nature puts out a percentage of homosexuals, to safeguard against overpopulation. It would still be a defective gene, but if nature did it for the sake of survival, it could be considered normal"

"Not bad," I answered. "But for that to be convincing, I would like to see some serious studies which support it."

"You mean like the studies that showed that right after a war, the percentage of male babies...skyrocketed, to make up for the war casualties?" Christopher asked.

"Precisely," I answered. "We need to study what happens when conditions in a given territory get crowded—whether the percentage of homosexuals increases."

"It could be tried with mice first." Christopher suggested. "It'd be interesting to see how...Lesbian mice perform."

"Normal or defective though," I continued, "the conclusion here is that homosexuality can not be a...*conscious* choice. These are the cards homosexuals have been dealt, by Nature. The rest of us have no business judging their lifestyle. Society is not threatened by it—though fanatics in some organized religions seem to imply it."

"The issue is not going away, though," Paul concluded. "Even if your theory is correct, problems still loom big in the foreseeable future."

"Give us a for instance," Christopher asked.

"Well, take their acceptance within the rest of society." Paul replied. "The explanation using genes may dampen the prejudice, but it won't completely erase it. Deep down we all feel resentment—it might be part of the gene...conspiracy, as you called it—because they don't contribute to the survival of the species. They do not—they can not—have...offspring."

"True, but this is already been addressed by...adoptions, artificial inseminations and the like," I asserted.

"That is another thing people resent," Paul said. "Their desire to bring up children in what is perceived as an aberrant environment. Most people cannot understand their desire to have children—they have professed no interest in...mating with the opposite sex. I have difficulty with this myself."

"Gays and Lesbians may have a defective sexuality gene," I replied, "but their procreation gene is usually intact—they want to contribute to the perpetuation of the species as much as you and I. You see how easy it is to explain these things, with the GSP theory?"

"Even if people understood their desire to have offspring, they may still object to having them raised in a gay environment," Paul countered

"They object, because they think that the environment will produce a...homosexual," I countered. "If the GSP theory is correct, that is not possible—some studies appear to confirm this, already. Anyway, I am going out on a second limb—here and now—with my prediction that such children will not turn out to be...gay—at any higher rate than the rest of society's children, anyway."

"OK, OK," Paul said. "Another serious future problem that I see, is this. What happens if they actually find these genes, and they turn out to be defective as you predict, and then, they find a therapy that converts them into...heterosexuals?"

"That would certainly create a...plethora of new problems, for them and society—especially moral ones," Andrew agreed. "Is the therapy...approved? Do they go through with the therapy? How does the rest of society—or even themselves—react toward those that do and those that don't?"

"I can already see all the self-proclaimed ethicists with their gloomy faces on round tables all over the world, showing their concern —never proposing any solutions, of course." I said. "I even see a new profession springing up among psychologists—homosexual gene counselors."

"But the world will survive even this...crisis," Christopher said, "just like it survived all these new discoveries of producing offspring with-out...physical contact. Are we through with this issue, I got some ques-tions on transvestitism and some of the sexual fetishes."

"No we are not!" Estelle answered...emphatically. "What happened to my arguments in the short episode "A Gay Life Style"? Your father here (she was pointing, but not looking at me) never countered them, then. Are you guys also going to ignore them, now?"

"Let's see." Christopher said, paging through the pages of the short episodes. "Here's the one on Gay Life Style, let's see. blah, blah, blah, hmmm...oh, here it is. All you said, Mom, was that you think it is a...choice—an alternative life style."

"And you guys are saying genes—and more genes. OK, let's see you explain...this. Our church runs a center for converting gays and Lesbians back to the...normal life style—on a volunteer sign-up basis, of course. Well, they have been having some success—some gays are returning to the...fold. Now that wouldn't be possible if it was...genes, would it?" She started looking out the window, pretending to be...disinterested. She was sure she'd scored, and scored well.

They all looked at me...smiling.

"Oh, not you again," I said, a bit irritated. "And just when we had this thing pretty well wrapped up."

"You thought you had it all wrapped up? Excuuuse me!" she said...sarcastically. "Should I withdraw the question so that you can continue to deceive yourselves that you have explained homosexuality?"

"Of course, not!" I said resolutely, "you know the rules—all questions are...game."

"Well, then?"

"How many people do they *claim to be able to*...cure? What percentage?" I asked, taking on the airs of an attorney questioning a...hostile witness.

"I think, I heard around 50% at the potluck the other night."

"You mean that 50% assume a heterosexual life style, not that they are really cured of their attraction to other men, right?

"I don't know,' she replied. "All I know is that they embrace the fact that homosexuality is a sin, and they abstain from having sex with other men. Some show it by getting married, some even go on to have children..."

"All right, but that is not a cure—only a change in behavior. No different than heterosexuals assuming a life of abstinence—priests, monks, nuns, etc. If they are still attracted to men, they remain gay."

"But since we don't know, and to exhaust the argument," I continued, "I will accept that 50% are actually...*cured,* even though I don't believe it. Now tell me, of those 50%, how many return to their old...gay life style, after some passage of time? How many? How many?" I kept pressuring the...witness.

"I heard that half of them go back to their...sinful ways," she answered, somewhat subdued.

"Which means, a 25%...apparent success. Now," I started pacing, looking at the ceiling, pretending that I was squeezing the last vestige of gray matter in my brain, "if we subtract another 5% who are lying—they went back in the...closet for obvious reasons—what have you got?"

"20% success rate," she said." "Not insignificant."

"That," I said with a look of...eureka in my face, "is not even as good as the...Placebo effect[9]."

"The Placebo...effect?" she said...aghast.

"Yes, why not?" I said. "You got a...therapy, you claim a cure—you got the Placebo effect—right in your midst. You know that—you are a nurse!!"

"Your Honor," I continued, pretending to be addressing some judge at the corner of the Den, "I rest my case," and I sat down, mopping my brow. Boy, I thought, that was...too close for comfort.

"And how do you explain the Placebo effect with your...gene theory," She asked—rising like the...phoenix from the ashes of the fireplace.

"Dad?" All eyes were on me again.

I had to think for a while. Estelle was a tough...adversary. I didn't really have an answer just then. So I quickly opted for postponement.

"Your Honor," I said again. "I'll have to take a...rain check on the explanation of the Placebo effect—it is not part of the group of genes we are presently discussing. We will leave it for the end—the Interview on the RAQ's (Rarely Asked Questions)."

"OK," Andrew said with a deep voice—imitating a judge. "Court is now...adjourned."

"Before we close this subject," Paul said, "I'd like to see you explain some of the other sexual disorders using genes and the survival principle—GSP."

"Like which?" I asked.

"Well, sexual fetishes, transvestitism, exhibitionism, coprolagnia, raping, and all the other sexual aberrations—whatever they may be."

---

[9] Placebo Effect is the reason given when patients are cured taking a drug they *think can cure them,* even though the drug is water or sugar pills. Up to 40% of patients exhibit the Placebo Effect depending on the disease and the...scientific study.

"All these are abnormalities—caused by defective genes," I said. "Detailed explanations are, of course, dubious, risky. Even so, I am willing to give it a...go. And everyone is invited to do the same. It can be a lot of fun, as long as we don't take these explanations too seriously."

"I got transvestitism right here," Estelle said. "His mother yearned for a girl but got a...boy. She dressed him up like a little girl when he was young, and this became an...obsession when he grew up."

"That would be an explanation using...psychological mumbo-jumbo—not GSP," Andrew said. "Let me give it a try. If the guy is a homosexual—that has already been discussed. Now take the guy who is a heterosexual, but likes to dress up in women's clothing. I am postulating the existence of two genes, which encourage us to show off our sex by our appearance. I said two genes, to take care of both masculine and feminine appearances. All of us have both these genes, with on and off switches, as usual. It is not hard to see, that if these switches are mixed up, you end up as a man, but desirous of appearing as a...woman."

"That was great!" I said. "I could hardly have done any better myself—even though I'm responsible for this here theory. Any other volunteers?"

"I think I can handle exhibitionism—now that Andrew defined this sex...appearance gene," Christopher said. "The ultimate in appearing to be a man is to show your genitals—your whole...naked body. If this gene is defective you throw out the...subtleties, you wear the coat—you...flash the goods."

"Excellent," I said. "Paul, how about you? You don't really have to accept the theory to play this game. You just need a bit of imagination in inventing some new...gene. If there are 140,000 of them, anything you dream up is likely to be in the bunch!"

"OK," Paul said. "Let me try masochism, a sexual aberration first described by some Austrian named Masoch in the 1800's—I looked this up recently in anticipation of this symposium. Personally I haven't met anybody with this... malady—getting sexual pleasure from being dominated, or even cruelly treated. But I'll take his word for it. My guess is that it's due to...cross-wiring two genes, one associated with the urge to be submissive to our superiors in society's hierarchical structure—the other a gene of sexual pleasure. I am also guessing that it is most common with male...homosexuals."

"Oh, really? How so?" I asked. "The explanation was excellent, by the way."

"Well, it's only a guess, mind you, but it seems that being men, they have strong genes for submission to authority—other men higher in the...pecking order. Now add their passion for having sex with men, throw in a little...defectiveness in all these genes, and you got the recipe for...masochism. Does it make much sense?"

"I think it does," I said. "I say, I never expected you guys to be so...adroit at this here game. Any other ideas? How about rape, or sadism?"

"I don't think those are very interesting," Christopher said. "They are also pretty easy to explain—violence genes gone defective, taking...down some sex-pleasure genes with them. Or procreating genes gone berserk...skipping the courtship stages and going straight to the...meat—impervious on whether the other person...wants it or not. I am more interested in the fetishes and the hard core abnormalities. How do you explain Adolph Hitler's abnormality, for example—he couldn't climax unless his partner...urinated on him. A form of coprolagnia I guess."

"I heard that some people can't reach their orgasm unless they are chained to a bedpost," Paul said. "Yet others need to be cuff-linked and whipped with a leather belt—otherwise no...go, or is it no...cum."

"Heck, that's nothing," Andrew said making sure Estelle had left the room to get more coffee. "I heard there are people who can't even get an erection without some...shoe up their rear. One guy—President Clinton's adviser, I think he was—had to get down on his knees and bark like a...dog before he could mount his partner, and then, only...dog style. And what about this famous sportscaster who used to...bite the girl's back, just to get himself in the mood for...mating?"

"I got a good one, too" Paul said. "I heard that a famous actor could not climax unless his partner...defecated on him—real coprolagnia—pretty disgusting. Yet, if Dad here is right, and it's due to a defective gene, you can't blame the poor guy—just pity him."

"That goes without saying, and it is part of the consequences of accepting the theory," I said. "These consequences include changes in attitudes—if the theory proves to be valid. Even changes in the penal code might be in order—we will see this at the proper juncture."

"We are now going to end the discussion of sexual abnormalities," I said, "even though improvising explanations for them, seems to give you guys some...kinky pleasure. Instead, I am going to finish this group of genes by playing my own...devil's advocate. I am going to ask a question, and then challenge you guys to come up with the proper explanation, using GSP, of course. Here it goes."

I got up, made sure Estelle was still preparing the second batch of hot coffee, and then laid out perhaps the toughest question of the whole symposium.

"Why is it that men like oral sex?" I said, making sure my voice wasn't strong enough to reach the kitchen.

Their eyes opened widely—they appeared astonished at the question—then pensive and reflective. I think I saw Andrew jotting down something on a napkin on the coffee table.

"Is it because President Clinton made it popular, and we have a gene which makes us emulate our leaders?" Christopher asked.

"NO, no, I mean genuinely like it—people who have never heard of President Clinton like it. The question is why did nature put in a gene, which makes men like it? Remember, according to GSP, if a trait appears in most people, the survival principle must be invoked."

"I know what he means," Andrew says. "We got to find how the pleasure gene for...fellatio, helps the species survive and flourish. If it does, I guess President Clinton should be elevated to a...hero—for lofty acts, beneficial to the species."

We all laughed. Funny stuff.

"Well does anyone have any ideas?" I insisted. Just then, Estelle was re-entering the den, with a fresh pot of...coffee.

"I got an idea," Andrew said. "It protects the species from overpopulation, by diverting our attention to a non-procreating orifice—reducing the odds for pregnancy. I wouldn't be surprised if there is a pleasure gene for every...orifice, for the same reason." Strain to come up with the proper words was written all over his face, when he saw Estelle re-entering the room.

"What are you guys talking about now?" Estelle asked curiously.

Everyone started looking out the window, admiring the lake view.

"How about, it provides variety in life, and said variety helps bind the family unit to a more cohesive entity—big help for the...species." It was Paul this time.

"I think I got the best answer right here," Christopher said. "I just read the other day—pure...happenstance—that the final secretions, contain some form of...antibiotic. It is now known that it is there, to purify their intended...destination—to keep it...free of infections. Now back in the old tribal days there were no...dentists, so the gene is there to provide for proper...dental hygiene for the recipient."

"You are becoming...incredible," I said.

"What on earth are you guys talking about? Why isn't anyone answering me? Are you guys scared I might raise some objection you can not answer?" Estelle asked.

"We were trying to resolve how nature intends us to keep our body parts free of infections," Andrew said. "I think we are done with it now—it wasn't very interesting, anyway."

"In that case," Estelle said, "I've got a question that needs explaining—it is quite apropos to male behavior—assuming, of course, you are capable of self analysis."

"We certainly are," Andrew said. "Fire away."

"Why are men always bragging about their sexual…prowess? I suppose your answer will be that there is some gene that compels them to do it—that is always your answer, anyway. But why should there be such a gene?"

"And, Dad," Paul added. "If the trait is universal among men, you'd better come up with an answer that demonstrates that this gene…contributes to the survival of the species"

"Right," I said. "Does anyone besides…the theorist want to take a crack at the answer?"

"I do," Christopher said. "First of all, Mom, the trait is hardly peculiar to men. I've seen plenty of women who brag about their…conquests, or their various…admirers as they used to call them in the old days. Granted they don't use…graphic descriptions of their encounters, as men do, but this maybe because society demands that they keep a…low profile. If Dad here is right, societal restrictions contrary to gene dictates cause plenty of…stress, so I am glad I am a man, and I can brag all over the place using any kind of language my…genes dictate."

"What about the gene?" Estelle asked. "Are you done with the preliminaries, so that you can explain the reason for the gene?"

"I have an explanation for the bragging gene, but Dad here will have to decide if it is acceptable or not—he is the major self appointed theoretician of his own theory. Here it is. The bragging is done to announce to the rest of the tribe that you are doing your major duty in life—contributing directly to the species. The genes are unaware that there are now ways to avoid pregnancy—they think that every sexual…encounter is for procreation."

"That is a fine explanation," I said. "Just exactly what a devout advocate of GSP would come up with."

"That would be a good explanation for bragging about sexual adventures if they were true," Estelle said. "Men's bragging, I understand, is based on…imaginary or.. wishful sexual activity."

"Well," Andrew said, "such a bragger maybe hoping for a…band wagon effect. If the word gets around he may actually start getting some…takers."

"Besides," I added, "bragging helps the person's image in the eyes of the tribe. The genes...demand that you do some...public relations about your activities, to win acceptance by the tribe. It is quite necessary for...your well-being."

"OK," Estelle said, "the explanation is quite believable. Actually I am starting to like this game of explaining human behavior, using Aris's GSP. I recall having a great time back in the 60's—doing the same thing with the principles of psychoanalysis. Do you recall that, Aris?" She asked me.

"Yes I do, indeed I do." I said. "It was widely practiced—and great fun. Of course it was much easier then—a lot easier than...GSP?"

"How so?" she asked, somewhat...annoyed.

"Well, we were trying to explain things, using two or three basic principles. Here we got some 140,000 genes to contend with. And you can't just wildly invent one to suit your needs it has to conform to the GSP principles. It has to be such that the survival of the species is enhanced by its presence."

"Which three basic principles?" she asked.

"You know, toilette training, Oedipus complex, fear of death, this kind of...rubbish," I said.

"Rubbish is it? She said. "And I suppose you think that this GSP is...God's gift to behavioral science—sorry, the Devil's or whatever—you don't believe in God, I know!" She looked quite agitated in stammering— she still liked...psychoanalysis. I realized that I had to...give her some minor victory—just to keep the peace.

"You know what I think?" she retorted. "I think the opposite is true. It's much harder to explain human behavior with only a few principles, than it is with 140,000 genes at your disposal. Do you want me to prove it to you?"

"Yes, go ahead," I said, seeing the possibility of an opening for a...peaceful conclusion.

"OK, ask me to explain any kind of human...quirk using this GSP thing. Mind you I am no expert on it. But I'll come up with an explanation—and in a jiffy." She said, looking quite confident.

"OK, since we are mopping up the sexual group now, how about explaining…fetishes—object fetishes? Why is it that some people can not enjoy sex without some specific…object in the vicinity, like a shoe for example? You might want to contrast your…GSP answer, with a…Freudian answer—just to see the difference." I said, using a mi d, agreeable tone of voice.

"OK, here goes. First a Freudian analysis, though I profess no expertise in it—can't vouch for its…authenticity. The person in question probably had his (or her) first sexual pleasure while some…shoe was in sight, and he (or she) got…fixated on it during the ultimate…bliss. Thereafter the shoe became necessary for the attaining of similar…pleasures. Well? How am I doing so far?"

"Sounds right, but I am no expert, either," I said. "The word…fixation gave it some sense of Freudian…verbosity, I surmise. Now the GSP explanation"

"OK, the guy (or gall) has a…superstition gene with strength 10, or maybe a defective one. You guys already admitted that such a gene exists—I don't have to invent my…own. Now this same person, has a wonderful first sexual encounter, and he is wearing or seeing some…shoe, while the ultimate bliss is reached. His aforementioned gene of superstition causes him to think that only in the presence of such a shoe, will he be fortunate enough to reach the same orgasmic pleasures. Thereupon you have the…shoe fetish."

She got up and started picking the coffee dishes—a sense of arrogance in her…strutting movements. I knew we had reached our…peaceful conclusion.

"You see," she says with a look of mild contempt, "it wasn't all that hard. The fact that there are all these thousands of genes, makes the explanations quite easy—you can invent a gene for …all occasions. Who is to say it doesn't exist, in this huge bag of 140,000 of them? At least Freud….."

"It won't be long before the whole DNA is deciphered, and we know exactly which genes are there and what they are meant to do," I said. "They estimate that this may be done by 2008, or so."

"I wouldn't be too...anxious to get to that point" she said. "All these explanations may turn out to be...rubbish—same fate you attributed to the Freudian ones. At least he enjoyed some fame for decades and still enjoying it, of course—present company excluded."

She had a point—even more than one. But I wasn't about to yield to a lot—just enough to pacify her and go on with the discussion without her renewed... antagonism.

"OK," I said, "you proved your point, you can use the GSP theory quite handily. But it isn't as easy as you make it out to be. You maybe as much of an...expert as anyone in this group. You have been hearing this stuff from me for decades now. Heck, you probably have the equivalent of a...Master's Degree in GSP, without ever paying a single dollar for...tuition. You should, at least, show some gratitude for all this free education you are getting." I concluded jokingly—no interest in starting her on any new...polemics.

"Putting up with you all these years has been my...tuition payment," she replied. "Pretty high tuition, too."

I didn't push it. There was no point—we had a lot of work to do yet, to finish the theory.

ARE YOU..."MENTAL"?

"I really like your title in this section." Estelle said, "It is pure nostalgia. I always look back fondly at the time we spent in England, during one of your Sabbaticals."

"Me too," I replied. "I still love some of the idiosyncrasies in their speech—still use them when I try to act...continental. Labo'ratry instead of laboratory, rubbish instead of...garbage, and my favorite in the title of this section—their use of the word...mental.[10]"

"But it's time to get back to our subject," I added, "the subgroup of...mental disorders is awaiting. My main thesis here is that so-called...psychological disorders, have their origin in physical defects in the brain—not some psyche, or a soul. These defects could be caused by a defective gene that gave the wrong... blueprint to that part of the brain, or from damage to the brain from other physical forces. In short these psychological maladies are actually as physical as a belly-ache—they are totally misnamed. And this goes for the phobias that we have already covered. There you have it—I told you we would eventually cover it!"

"So," I said, concluding my short introduction, "with my contention that defective genes, proteins, chemical/electrical or other types of physical brain alterations cause these disorders—who has the first question? I know you are all...itching to attack."

"So you are still staying with the same...song and dance routine—the GSP—even for mental disorders?" Paul asked. "These were thought to be mostly...psychological for centuries, you know."

---

[10] Mental, in England means...crazy

"Yeah, Dad," Andrew said. "Even when I went to College the course was called Abnormal Psychology—you are up against years of conventional wisdom here—not to mention thinkers, who are...giants in the field of the psyche."

"Ill take my chances, I tell you," I said. "The fact that the name was changed from Abnormal Psychology to...Mental Disorders shows me that the trend is going my way."

"I agree with Dad," Christopher said. "The name change heralds a new recognition that the problem lies in the...brain and not in the...psyche. Once, I asked our professor of Psychology 1 to tell us where the...psyche is. He immediately became flustered, embarrassed—red in the face. He started talking about people who have...heart and soul, an inner self where the...ego resides—all kinds of BS, till the...bell rang."

"I think the Psyche is still ...missing—nobody has found it yet—it might be hiding somewhere inside the brain" I said. "If that be the case, if we agree that all these problems are mental, if there is no...psyche anywhere to be found, what in heavens is a psychological problem? What is a psychological factor—a psychological *trauma*?"

"I think those things have to do with people's...feeling," Estelle said. "You guys wouldn't know much about that—you are all on an ego trip, trying to explain everything with chemical or electrical reactions."

"So feelings are not caused by chemical or electrical reactions in the brain?" I asked her.

"I don't think so," she said.

"Can you be more specific with examples?" I asked.

"I got some examples," Paul said. "What about finding out that your child is seriously ill? What about witnessing a murder, a massacre? What if someone calls you an...imbecile?"

"Or conversely, " Estelle added, "what about finding out you child excelled in school, your team won, you got a promotion, or won the...lotto? Aren't all those reasons for changing your feelings, your entire...mental make-up?"

"Certainly they are," I said "I am afraid you are confusing feelings with the…body's mechanism of causing these feelings. I have never denied that feelings exist—the GSP is not some…heartless theory as you are suggesting. People are emotional beings—they feel anger and happiness, and all the rest. All I am saying is that…genes set the character traits of the person, and within these bounds the brain processes information, judges it, and causes these feelings. Of course, the issue is more complicated than that—but that's a simplified way to explain it. The main idea here is that it is chemical alterations and electrical activity in the brain that cause…sadness, and not some…psyche that feels hurt."

"And how do they know when to cause happiness or…despair?" Estelle said.

"Look, I already told you, the issue is extremely complex. We can simplify it, somewhat, by invoking our GSP theory. We can divide feelings into two groups Those that deal with…good feeling and those with…bad. Experiences, that help the tribe survive and flourish, always result in good feelings—the opposite being true for bad ones. In between there are many shades or feelings that the brain causes, depending on how the information received was judged. If the information received from the…happening is not…clear-cut, if the brain is…puzzled, so are we—we don't know what to feel. The key judgement appears to be whether what happened aids in the survival of the tribe/species or not."

"So you are saying that when we witness a murder—bad for our tribe—the brain causes the production of a chemical which makes us feel terrified and sad? Is that right?" Andrew asked.

"Absolutely!" I said. "It is pre-programmed to do so. And if we witness an event that helps the species—two lovers kissing, say—then the produced chemicals cause joy, happiness, contentment, even exhilaration. In between these two extremes you have thousands of genes, chemicals and brain spots, specializing in other emotions. I'm not going to list them all—the list would be enormous. However, many of these emotions are outlined as character traits in the personality genes—the last group in the theory. Now how strong these emotions are, depends on the strength of the gene for that particular trait in your character."

"To put it bluntly," I went on, "if you see a naked...babe you are programmed to feel...lust—among other emotions. Now how much lust you feel depends on your...lust gene—not some psyche lurking in the shadows. Of course, you are not in control. It is all part of the gene...conspiracy. And I don't expect you to agree with me—you can't."

"And what...proof do you have that all our feeling are caused by chemical or electrical activity in the...brain?" Estelle said, looking sure that the question would...throw me.

"There is plenty of evidence that this is so—some of it there for decades," I replied. "Take alcohol, nicotine, other drugs. These are chemicals that produce mood changes and feelings, when they hit the brain. If a chemical can give you a feeling of...happiness, that should have alerted the psychological community long ago that they were on the wrong path, searching for some psyche."

"It was in yesterday's science section of the paper that scientists have detected increased electrical signal activity, just before we start feeling some emotion. Even though I am not in Dad's camp yet, I have to be truthful, don't I." Andrew threw in, making a statement rather than asking the last question.

"I read, pretty recently," Christopher volunteered, "that science discovered the reason...for addiction. We get hooked on tobacco because its nicotine causes a substance to hit the brain—a chemical that gives you a feeling of contentment and...well being. I just heard on the news that some pill was found that eliminates this effect, and helps you quit...smoking. That is more proof of the relationship between chemistry and feelings, isn't it?" he asked.

"It certainly is," I replied. "And tobacco addiction was one of the...biggies for psychoanalysts. It wasn't long ago that we thought it was...a psychological addiction, having to do with...manhood, role model imitation, family environment, etc. I even recall some analyst who claimed that the problem was an addiction to a hand…ritual—pull out the pack, take out the cigarette, light the cigarette, put it in your mouth. His therapy consisted in…destroying this hand ritual, by finding alternate ways to…occupy your hands during your waking hours. Good stuff!"

"Same goes for alcohol addiction." Christopher added. "The reasons previously given for that varied from domineering parents to…escaping life's hard realities and heavy burdens. Now we know it is related to a defective gene, and it appears to be in a hereditary cycle."

"OK," I interjected. "Let's just look at some of the…mental disorders under discussion. How many of them are…cured with the…couch treatment—searching for childhood traumas to find their cause? I put it to you, their causes are chemical or electrical imbalances in the brain, caused by defective genes or other…physical factors. That is why the only treatments that work are…pills—good chemicals to…neutralize the bad ones. I believe they are called…psychotropic drugs—the last hurrah for the "psycho" prefix."

"Dad is certainly right about Tourette's syndrome, the mental disorder he devoted a short episode on," Andrew said. "I would like to add that he is also right about…epilepsy, though I don't see it in the list—it's OK, he said the list isn't complete. It wasn't long ago that epilepsy was thought to be a…psychological disorder, but it proved to be caused by electrical imbalances in the brain—physical factors, of course. The churches claimed a demon was inside the afflicted causing the fits, for reasons only the demon seemed to know—no motive was ever provided."

"The churches love any disorder they can attribute to evil spirits or the…devil—it proves that the devil exists," I said. "If the devil exists—who can dispute…God's existence? Things always come in pairs, you know. News that science found physical reasons for such disorders is not only cause for…consternation in psychological circles, but in… religious circles, as well."

"I'm in Dad's corner on mental disorders," Paul said, "though I am still wavering on the overall GSP theory. I just read on the Sept. 27, '99 issue of Newsweek that scientist have recently found that schizophrenia may be caused by poor fetus diet, during the early stages of gestation. That is a long way from Oedipus complexes or psychological traumas in childhood. Besides, I know several people who find their relief for their mental depression at the pharmacy—pills—and not at some psychologist's or psychiatrist's office. The only reason they visit the psychiatrist is to get the prescription renewed—not to get psychoanalyzed."

"Can I have the last question, before you move on?" Estelle asked.

"Yes, certainly," I said, none too happy about it, but never showing it.

"Is there no chance that some...shocking event—I won't call it psychological, since you appear keen in...*wasting* the word—can alter your mental state and send you into mental depression or worse?"

"I suppose there is," I said, "I have never denied that traumatic events can cause mental disorders. All I am saying is that they do so by a process of affecting the physical well-being of the brain—that they are a biological problem, not a psychological one."

"Can you illustrate this with an example?"

"By all means," I responded. "I'll be glad to give you a typical scenario, as I see it. Let's say that a mother witnesses the murder of her child. Her brain judges the event bad for the tribe's survival, and horrendous for her—her child was her direct contribution to the tribe's survival. Proteins or other chemicals blueprinted by her genes rush to her brain to cause her to feel shock, despair, helplessness, anger. Other chemicals also rush there to calm her down—a pre-wired response, also dictated by...genes. But what if one of the needed chemical is no longer produced, because some gland doesn't work right anymore? What happens if one of the chemicals is now secreted at dosages larger than normal? Her brain may suffer damage—she may become clinically depressed. Her hair may even turn...gray. Her main hope after that is not a...grief councilor, but some chemical—both for her depression and for her...hair."

"So besides the depression—a mental disorder—she also get a prematurely white set of hair. This leads to a loss of self-respect—intensifying the depression," Christopher added, frowning.

"I think this whole discussion is driving me...mental," Estelle said. "Your ideas are too iconoclastic—against all conventional wisdom and traditional thinking. I've got a headache coming, I'm going in for a nap, before these chemicals totally flood my brain and lead me to some...nervous breakdown."

She got up and started for the bedroom. I said nothing. After years of living together, I realized that this wasn't a concession speech. Even so, it was a silent admission that she had run out of...arguments. I chose to consider it more than that—a resignation to the inevitable. I relaxed, worked myself into a more comfortable position in the chair—started thinking of the next topic.

"Dad," I heard Andrew calling out, jarring me from a sense of pious complacency. "Did I just hear you admit that under certain conditions— or life experiences —one could end up with a mental disorder?

"It happens rarely, under extraordinary conditions similar to those I described to Estelle a minute ago," I replied. "And it is caused by brain or gene damage—always."

"Is that so?" Andrew replied, looking unusually revived, "Isn't that a contradiction in your theory? If such dramatic events could effect your psychological—sorry, mental—state, why are you ridiculing traditional psychology for making the same claim?"

"Yeah, Dad," Paul butted in. "And if such an event could cause the disorder, why couldn't recalling the same event—via the use of the...couch treatment you so despise—reverse the damage? I think your theory here...validates the theories you dispute—the opposite result of what you intended." He got up, walked to the door and started shouting.

"Mom, come back. We got Dad...cornered. Come watch him squirm—it's quite a sight!"

I heard footsteps and Estelle re-appeared, magically refreshed and ready to go to combat. I let Paul and Andrew bring her up to date on what she missed—it gave me an opportunity to prepare my rebuttal.

"Well, Dad" Paul shrieked. "What is your answer? Are you admitting that all your past criticisms of psychoanalytic thought were misplaced?"

"We are not saying that your GSP is wrong, mind you" Andrew said, "just that it...augments all previous thought—puts it in a new perspective," he added sarcastically—a patronizing look in his face.

"We'll see," I replied. "Would you allow me to reverse roles with you guys, becoming the...devil's advocate myself—against your new arguments?"

"Certainly," they both replied.

"OK," I started. "Let's take the following scenario. A guy with suicidal tendencies—I just read in the Internet that a gene appears to cause them, when defective—decides to visit an analyst. The minute he steps inside the analyst's office, he has accepted an assumption about the origin of his mental affliction, has he not?"

"What assumption?" Andrew asked.

"The assumption that some past traumatic event has caused it" I replied. "Right?"

"Right," Andrew replied.

"How relevant is that assumption?" I asked. "Considering recent discoveries of the causes of mental disorders, where do you put the probability that this assumption is right? An inherited, defective gene could have caused his suicidal tendencies. They might have been caused by some illness his mother had when he was in the womb, by alterations in his brain from physical factors when he was an infant—all the things that we have discussed before. Right?"

"Yes, but also by some traumatic event in his childhood—you just have admitted it." Paul interjected.

"That too," I agreed. "But with a rather low probability—considering the plethora of possible causes. Am I right?"

"Yes, but the possibility is still there that the assumption is right!" Andrew insisted.

"Fine," I conceded. "But let us put it in some perspective. How does it differ from a pathologist assuming that your chronic fever is caused by a...spider bite? It could have been—of course—but with very low probability. Would you be happy with that assumption? Would you then agree to devote all your time searching for the...spider that caused the fever, hoping the antidote exists? Or would you rush to see another doctor?"

"I see your point," Andrew said. "But as low as the probability might be, the cause of the mental affliction could still be a past traumatic event—never mind that we now accept that the event caused physical damage, not...psychological. So, visiting the analyst would not be a bad idea in that case? He might uncover it, force you to...spit it out, and cure you, right?"

I started laughing—I couldn't stop myself—knowing fully well that I looked...arrogant. I quickly apologized for the outburst, and moved on.

"Well, let's concentrate on that for a while," I replied. "The scene shifts to a guy who suffers from severe depression. The assumption is that this depression had actually been caused by a childhood trauma; he watched his drunken father rape his little sister, and then barely escaped, when the father went after him with a butcher knife. The chemicals that flowed under this horrendous event caused serious damage to his brain—he is chronically depressed. He has totally suppressed the event—it is too painful to recall. Is that the scenario you want?"

"Perfect," Andrew replied. "Now tell me what would be wrong with seeking an analyst who could dig the event out of his subconscious, make him relive it—face up to it as they say. Would that not be helpful? Might it not even cure him?"

"Let us go on slowly, without jumping to any conclusions," I replied. "What you just concluded, is the analyst's argument—one of the justifications for his profession."

"What makes you think that reliving the repressed event will be helpful?" I inquired.

Andrew looked around for help—mainly to Paul and Estelle. They were both silent—appearing to be racking their brains for ideas.

"The conventional wisdom is that if you uncover and bring forth the event that caused it—it's helpful," Paul volunteered. "I guess people who went through it, report that it improved their condition."

"Never mind...conventional wisdom and anecdotal evidence," I retorted. "I want your opinion in the context of our discussion and the conclusions we have reached. If the event had caused physical damage to the brain—we agreed on this already—how would the reliving of it...cure the damage?"

"It would do it by causing the secretion of chemicals that can do it," Andrew replied. "After all, we have agreed that the event was very traumatic. Reliving it could still re-jar the guy to secreting chemicals all over again—same mechanism as before."

"Aha," I said. "So if you live a trauma once—chemicals pour out and damage part of your brain. But if you live it a second time, the chemicals are now beneficial? This time they miraculously neutralize the previous damage? Does that sound plausible?"

"I got to admit, that the way you put it, it doesn't," Andrew replied. "It is more logical to conclude that the same event would cause the same chemicals to be secreted. Could that, then, cause additional damage?"

"It might," I replied, "In fact, I believe that this has happened. People were forced to relive horrid events in their past, and their mental condition worsened—much to the chagrin of the attending analysts. The assumption that reliving such events cures mental damage is not only dubious—it might even be...dangerous," I concluded.

"So you don't believe that...facing up to such events, helps?" Paul asked.

"I believe that the phrase "facing up" is psycho-babble, in this context," I replied. "I think there is a better chance of altering the brain structure with...electroshock than with rehashing some old...trauma.

But electroshock is also dangerous, since we don't understand, nor can we control its effects on the brain—not yet, anyway. Still, I think electroshock originated with behavioral pioneers who sensed that to change behavior, you must concentrate on the brain, not the psyche. The real believers in this concept carried it even further—they tried…lobotomy, with the known disastrous results."

"I got an idea," Christopher jumped up, after a long silence. "What the patient needs is not to relive the old horrid event, but to relive some previous happy—exhilarating event. That would have a better chance of secreting the…opposite kind of chemicals—chemicals which cause the damage to be corrected—reversed. I hope you write that down in the book, so that I have proof that I thought of it…first. Don't be surprised if it soon turns out to be the latest craze in…mental treatment. It might turn out to be successful."

"I doubt it," I replied, "though I agree that the chances of success are better than reliving the traumatic event."

"In fact," Christopher came back again, "I don't even think that the treatment needs to consist of re-living a past happy event. It could consist of living such an event now, or in the future. If it is powerful enough, the secreted chemicals may be beneficial."

"Such an idea is not totally ridiculous," I inserted. "There are people whose mental conditions have improved by extraordinarily happy events—a passionate love affair, for example. Love is so entwined with survival and the…genal condition—the chemical secreted may be powerful enough to even cure mental disorders."

"Hey, is that an idea for a new business?" Andrew asked. "How about providing powerfully happy events for the mentally afflicted? It could help their brain or gene damage—no matter how this damage originated."

"It is a long shot," I replied. "Not to mention that it is off our subject. I think there is a better chance for finding beneficial chemicals via biological research than trying to find life events that might…secrete them. Looking for these events and staging them could prove financially prohibitive—a lot more expensive than swallowing a pill, anyway. Could we now close this issue?"

"So what is the final conclusion?" Estelle asked.

"Here it is," I replied. "There seems to be only a very small probability that a mental disorder is caused by a childhood traumatic event. Even if it is, the notion that it would be beneficial to relive it, appears to be as sound as the notion that it is detrimental to do so. In other words, even though I admit that enormously shocking events can cause brain changes and thus personality changes, the mechanism is not scientifically understood, and certainly not reversible by re-staging the same event."

"And what about all this anecdotal evidence that psychological therapies work? Are you disputing that some people with mild depression, phobias, panic attacks, anxiety, etc. have been cured by couch-type treatments?"

"No, I am not!" I replied.

"What?" Estelle shouted, totally puzzled. "I can't figure out what it is you are asserting. With one side of your mouth you ridicule conventional psychology, and with the other side you admit that their methods work. Is that not a flagrant contradiction?"

"No, it is not!" I quickly responded. "I am accepting that some people are cured by these methods, but at the same time, I am contenting that these methods are...*scientifically* groundless. If you find this statement contradictory, it is because you are ignoring the...Placebo effect."

"What? The Placebo effect again?" Estelle coughed. "You used this excuse already, in dismissing our Church's success with...curing homosexuals. If I recall correctly, you wriggled out then, of my demand to elaborate. Would you care to do so at this time also?"

"Wriggled out? I simply said that we shall reconsider the issue in detail at some later time. I was—you see—anticipating this conversation. I am ready to elaborate right here and now."

"Well, we are all waiting," Estelle replied, re-adjusting her chair's cushions. "Considering that you use the Placebo effect whenever your theory fails you, this ought to be quite interesting."

"OK," I started, "I think we all know the rudiments of the Placebo effect, but I will briefly review them, anyway. Sick people sometimes get well by taking a drug *they believe* *can cure them*, even though the drug is just a…sugar pill. When this happens, medical science attributes their recovery to the Placebo effect. The necessary item for the Placebo effect to work is—by all accounts—a strong belief on the part of the patient, that the drug he is taking can cure him."

At that point I stood up and turned to Estelle.

"Now I can explain what seems to you unexplainable—that I accept that people can get cured by psychological treatments, and at the same time, I consider these therapies useless—no different than quackery. This includes your church's program of homosexual conversion."

"I am dying to hear it."

"It is quite elementary—no different that in the case of medications. A treatment can only be scientifically acceptable—just like a pill—if it cures more patients that a sugar pill does, i.e. than the Placebo effect. That it will cure *some* patients—your anecdotal evidence—is a given, it is part of the Placebo effect. Even pseudo-doctors—quacks—can do that. In fact, such people have been running cure-scams for generations, with the Placebo effect always on their side. But can it do better than that? Can it consistently cure more that 40-45% of the patients? I personally dispute that. This is the sense in which I simultaneously accept the anecdotal evidence of some cures, and reject these therapies as…smoke and mirrors."

"I see," Estelle replied thoughtfully. Still, she would not let go—she had to take one final try. "And how, pray, do you explain the Placebo effect with your GSP theory?"

"That's another issue altogether, and this is the second time you have brought it up." I replied. "But it is still not pertinent to the present conversation. I plan to take it up later, at the end of the book—as an answer to one of the Rarely Asked Questions (RAQ's)."

Estelle left the room again, not saying another word—looking exhausted. I was actually happy that she wasn't going to be around for the final round of our symposium. I was anticipating a tough exchange—a possible...waterloo for the theory. Good thing the boys were also getting tired—I thought to myself. Fatigue tends to make people a lot more...agreeable.

THE PERSONALITY GROUP—A WATERLOO?

"If there is to be a...show stopper for the GSP, it's this," I started, "the individual personality group. Who wants to take the first...whack at it?'

"Let me start," Paul said. "since I still feel somewhat uncomfortable with this GSP theory. I got to confess that up to now most of my questions have been answered. So I am coming around, all be it, somewhat slowly."

"My first question is this," he continued. "Even if the theory is correct, has nature gone that...detailed? Are you sure that we have a gene for...neatness, one for congeniality, one for... pusillanimity—whatever that is[11]."

"I am not sure," I said. "but if we have 140,000 genes—so anything goes. As for my list, I just put there any character trait that I could think about. Some of them are similar—they are synonyms—or their meanings have some...overlap. Others are antonyms, especially when you go from the positive ones to the negative ones. If we revoke the idea of the strength of the genes, then a trait and its antonym may come from the same gene with diametrically opposite values."

"What Dad means," Christopher volunteered, "is that generous and miserly, for example, may be the same gene—normal strength around 5. The miserly fellow is sitting at 0 or 1—the generous one at 9 or 10."

"Right," I said. "Note also that some of these traits appeared in previous lists."

"Yes, I saw that," Andrew said. "Brave and violent were in the Survival Group under Life and Territory, for example."

---

[11] Pusillanimity is cowardliness—Webster's Dictionary.

"Exactly," I said. "That was because it was easier to explain then, why we have them. However, they are still character genes, so I included them, as well. Anyway, the list is only there to help us...talk the talk, and...walk the walk. No pretense of completeness is made here. The basic contention remains that all character traits—whether I can identify them or not— have a gene as their origin. And don't forget their strength either—it will play a part in the explanations."

"Of course, a typical individual does not accept this concept," I continued. "He feels that he himself was responsible for adding a given trait to his personality. If he is brave, he is sure that he...made himself brave—by some conscious decision to be so. It is all part of the...gene conspiracy— how else can I put it? It is that...miserable cover-up gene that incapacitates you, and dishes you the...illusion that you are in control."

"I really liked the I'll have some Vanilla episode—probably because I know Antoine personally," Paul said. "Even so, I am still finding hard to accept that this type of trait comes from a gene—that it is not...learned. Acceptance of such a premise would mean that we are actually...stuck with a personality—that we have no choice in life. In some ways, this theory is a return to some old ideas of a...predetermined destiny—kismet or karma—though those theories attributed it to a path laid out by some...God."

"Yeah, Dad, you may be overdoing it claiming that there are genes for such minor traits as...congeniality," Christopher said. "I'm with you on some of our hard traits—those needed for survival—but keeping a...chronic smile on your face is something you can learn."

"Especially if your mother is keen in entering you in some beauty contest," Andrew added. "The congeniality price is a real...biggy. But I got to question this insistence on genes for mild character traits like...enigmatic for example. Heck, men will learn to do anything in order to attract babes—lie, cheat, even be...enigmatic, if the situation calls for it."

"And that's not the only...incentive in life either," Paul added. "Even...dogs learned when the incentives were there—I'm sure you know of the Pavlov[12] experiments. I have a feeling that you are wrong here. I think that a person's character evolves as he goes through life—he learns by experience and adjusts accordingly."

"I believe that a person's character might also change instantaneously." Andrew said. "What about all these cases we hear about—mean characters turning into…angels, because they stared death, right in the face?"

"OK, guys," Christopher said…coming to my rescue. "Let's have a pause in this…bombardment. Let's give Dad some time to answer all these questions, before we fire some new ones. After all, he is getting on— he may not be able to…remember them all."

"Thank you Christopher," I replied, "I'm sure glad you are on my side, even though it barely…shows. Let's see if I can put some order in this chaos."

"Let me start with the last question first—the one about guys turning into angels because of a…near death experience. I've heard some of those stories myself—they are quite popular as…movie scenarios—real…tear-jerkers. The trouble is that they never follow these guys long enough after the…conversion, to see what happens next."

"Are you saying the stories are…fake?" Paul asked.

"Not exactly," I replied. "I am saying these stories don't end there— you've got to follow them through before you draw any conclusions. If the guy died, he is not of interest in this discussion. He did his little…conversion routine for reasons of his own, but he is not a verifiable case of an…altered personality—just…dead, hopefully in his…happy place."

"But if the guy lived on, did he remain an angel, or did he revert back to his old, mean self, as soon as he saw that his imminent death was…exaggerated? This is what must be studied before we conclude that a personality can change from some…traumatic event. My guess is that as soon as he sensed that it was a…false alarm, his genes…slapped him right

---

[12] Pavlov, a Russian, conditioned dogs to do a lot of things, using hunger, pain and other incentives..

back to his old mean self—even...meaner to make up for lost time. That has been my experience with such cases. The Hollywood stories end with the emotional highs of the conversion—bad guy repents, good wins over...evil—relatives and acquaintances are all amazed at the miracle. They don't bother to follow it through for a few months—it doesn't guarantee a happy ending."

"Now that I think of it," Andrew interjected, "Dad could be right. I know a few people who thought they were dying of cancer, myself. They all changed their character—became nicer guys—but not for long. As soon as they discovered it was a wrong diagnosis, they switched right back, to their old...petty selves. Only one of them retained the changes, but he looked kinda...funny—he may have suffered some brain damage as well."

"Look here," I said, "If the brain comes out altered from the experience, you could come out a different person—even a...vegetable. That, we accept as part of the theory."

"OK, OK," Paul said, "maybe the overall personality appears the same if no chemical or electrical alterations took place in the brain. But what about changes in individual traits due to...learning experiences?"

"Let me have a turn at Dad," Andrew said. "I got the following...scenario. This gal, call her Valerie, is very shy, almost anti-social—gene-wise, of course. She hates to be among people—hates to talk to people. But she needs a job to survive, and the only one she finds is as a...hostess in a busy restaurant. Well? Would she not change her personality to...survive? Weren't you the one who was telling us one day, that humans would do anything to survive—even...eat one of their own kind?"

"I heard that if you ever ate...human meat," Christopher said, "you'd love it—you'd become addicted to it. That's why they kill any animal that eats a...human. The beast becomes a...junkie—she has to be wasted."

"I appreciate your continued support," I said to Christopher, "but you are supposed to...change the subject, only if I am...sweating, and then only...subtly—not so...flagrantly."

"He's just kidding," Christopher said. "There is no such agreement. Dad—tell the truth now."

"I was only kidding," I said.

"OK, then," Andrew said. "What about my scenario?"

"You mean Valerie? Well, if she must, she will play along—be social, a gun-ho…extrovert." I replied. "But she won't be too happy about it."

"What's happiness got to do with it?" Andrew asked. "Is Valerie going to change her character because of the compelling requirement of the job or not? That's what we are arguing about—not happiness!"

"You'll see where happiness comes, in a minute. No, Valerie will not change her personality. However, she will change her behavior for as long as she is on the job—she will suffer through an acting job of being sociable, in order to…survive. But behavior against one's…genes causes stress—this is where the happiness part comes in Valerie will continue to work with an unbearable stress—no different than the rest of us, of course. Very few people get a job that totally…matches their genes."

"Is that a new definition of stress?" Andrew says. "If so, I never heard of it before."

"Well, it is not really a definition. I'm pinpointing the major reason for it—genes. There is some stress when one is placed in a situation he feels…incompetent to tackle, but that will go away as familiarity sets in. The main reason, though, for stress is the…mis-match with your personality genes. Can you imagine being a coward or pacifist working as a…bouncer in a bar? What can be worse than that?"

"How about an effusive…extrovert who is in prison and is placed in isolation?" Andrew asked

"I got a worse case than that," Christopher says. "A heterosexual in the island of…Lesbos."

"Seriously, though," he continues, "Would that explain why people claim to be highly stressed, even when they do ridiculously simple jobs in some…assembly line?"

"Precisely," I said. "Very few personality genes are pacified by seeing you put the same nut in the same...fender—always confined to the same spot of the floor. That's why coffee breaks were negotiated—people's genes were going...nuts with this type of jobs."

There was a bit of silence for a minute, so I got up and tried to summarize the section's conclusions.

"Here's what we have so far," I continued. "I am contending that personality—in its most intricate variations—is dictated by genes; that the strengths of these genes determine the variations in the traits; and that these things can change by physiological forces only."

"A person," I went on, "can make adjustments in his life—go against his genes, so to speak—especially if the adjustments are minor. If there is a huge gap between the...demands of his personality genes, and the life he pursues—the stress may be too high a burden to carry. He may suffer exhaustion, a... burnout—even a...mental breakdown. The stress could effect his brain...physiologically."

"I bet that's why everyone dreams of moving to a mountaintop, away from it all," Christopher said. "He thinks it is the only place where his genes will be at ease with the world."

"And he is wrong, of course," I retorted. "Some character genes will be happy there, but others won't—new stress will sets in, and in a hurry. The wise person knows...instinctively, that there is no perfect environment—too many genes to...keep satisfied."

"Even so, people will continue to fantasize about such a...happy place." Christopher added. "A place where the environment is in equilibrium with the...strengths of the genes. Personally I plan to adopt it as my new definition of...Paradise."

"Not a bad idea—thank you Christopher," I said. "Now in the absence of Christopher's...Paradise, most of us are dreaming of being our...own boss—the boss is always blamed, not the gene...mismatches. Or, looking forward to...retirement. They aren't quite the...ideal situation, but we don't know that—not till we...get there."

"You know something?" Paul says. "If Paradise is as Christopher defines it, I can see how difficult it would be—even for God—to create one. Since we all have a different...gene equilibrium, we all need a different type of Paradise. The logistics would be impossible even for God—everyone would expect a custom-made heaven. Dad here, for example, would love to be in a place, where he could perpetually...argue."

"You couldn't put him in a place with other...arguers, either." Andrew added. "He might occasionally...lose an argument and be unhappy. It would have to be a place where...volunteer angels argue with him, but always...lose the argument—agree with him at the end. That would be Dad's...heaven."

"It is exactly as you say—you may have come up with the reasons why Paradise is...impossible," I said, pleased with the results of the conversation. I was—unexpectedly—collecting material for my other book—on...religion.

"Don't worry about Dad," Christopher volunteered. "He found his own Paradise right here on earth—in these...books he is writing. He puts in there whomever he wants as...arguing partner—squelches them mercilessly, with no remorse. If you think this conversation will be reported in the book in an...unedited form—you are...fools."

"Well." I said, "I think this section is pretty much in the bag. All I have to do is...edit it—to keep my...genes in equilibrium, as Christopher said. Are there any final...jabs at the theory?"

"I have a general type question that concerns the overall GSP theory," Paul said.

"Let's hear it then," I said turning to Paul. "Since it pertains to the overall GSP theory I'll make it the title of the next section."

"OK," Paul said. "Here it goes."

"Do We Have a Choice?"

"It depends on the type of choice you want," I answered.

"What types of choices are there?" Paul asked.

"Two major ones," I said. "The choice to alter your personality—your character traits—and the choice of decisions and actions in your life—free will as it is often called."

"All right," Paul said. "Let's take the first choice. Do we have one, or is it all pre-determined by the genes?"

"Hey," Christopher said, "where have you been? That's what we have been talking about in the section we just finished, isn't that right, Dad?"

"That's right," I answered. "And the conclusion was that you don't have much choice there—you got to live with whatever character traits your genes...served you. You have the choice to deviate from the traits somewhat, if the situation demands it, but if you do, expect...angst and unhappiness. The key thing to understand is that you are not changing the...character traits—just your behavior. That is why you get stressed."

"So the conclusion is that we have no choice in changing our personality? None whatsoever?" Paul asked with dismay.

"If the theory is right, a categorical NO, we don't," I replied.

"Wait a minute," Christopher said. "That is not altogether right, even according to the GSP theory. You can change it by external forces that change the genes, or the parts of the brain that houses the character traits."

'Oh, yes, sure," I remarked. "You can start taking drugs that effect the brain. Or you can hang around nuclear power plants all day. Those things might change your genes and your personality, but you still have no real...choice, you can't control the results."

"Or you can wait till science can intervene in your genes and create a...custom ordered personality," Christopher said.

"OK, OK," Paul said. "I got the point. I'm glad I am not subscribing to this theory, myself—it is rather depressing. I always felt that you could improve on some of your faults with some effort—it was a goal of mine, especially when I made my list of…New Year's resolutions."

"Well, all you got to do is reject the theory and go on doing the same thing on new year's eve every year," Christopher volunteered. "If the…list is the same from year to year, your efforts to change your personality failed—Dad's theory here is vindicated."

"Exactly," I said. "That is a good experiment to test the theory! And, by the way, don't feel too bad about not accepting the theory—the genes don't want you to, anyway—they want you, instead, to think you are in control."

"Now, let us turn to the second type of choices we make in life—choices in our actions," I continued. "Can you guys help out on this? What do you think the GSP theory dictates?"

"I bet I know what happens when we live under…peril—the genal condition as you called it." Andrew said.

"Oh, yeah," Paul said. "I do too. That is the…zombie like state when the genes have taken hold of your brain and dictate your decisions and actions. Is that right?"

"Absolutely," I said. "When under the genal condition, you can forget about…free will. The survival of self and species can not…allow the use of your brain—we have repeatedly discussed this, it has almost become a…mantra. But what about normal conditions when the brain is in control? Do we then have any choices?"

"I'll take a chance," Christopher said. "I think the theory allows for such choices—subject to some restrictions. If we look at the Antoine episode—the conservative professor with the same shirt—he had a choice on…which shirt to buy. There are a lot of…conservative looking shirts, and that was the choice in this case."

"You hit the nail right on the head," I said. "We have choices of action in our life, but they are subject to the constraints put on us by our personality—the character genes. Going far away from those constraints, is also possible, but expect stress to set in quite rapidly."

"That is a great way of putting it," said Christopher, cheering me on. "I think you can…enunciate it as a Principle of the GSP theory. Choices of action are within the constraints imposed by the character genes. Sounds…scientifically profound."

"Can we have some examples to illustrate this…principle," asked Paul with sarcasm in his voice.

"Certainly," I replied, pretending I didn't notice the sarcasm. "Bear in mind that we are talking about normal…not defective genes—those with defective ones could have even more restrictions in their actions."

"Let me take our familiar personality, our Professor Antoine, of the I'll Have Some Vanilla episode. Do you know how he spends his vacation every year?"

"Staying home reading conservative editorials in…right wing magazines?" Paul asked.

"Not quite," I replied. "He goes to Williamsburg, Virginia. He stays at the same hotel—even the same room, I think—eats the same dinner at the same restaurants, walks around looking at the colonial stores in the reconstructed village. Year after year, the same vacation. Theoretically he has a choice—he can go anywhere he pleases. Practically he is limited by his conservative gene—his distaste for change. Do you see that his personality places constraints on his decisions?"

"I guess so," Paul said.

"I got another example" Christopher said. "Adrian, across the street. He has a very strong religiosity gene. But do you know what religion he picked? Rastaferianism! Don't you think that matches his flamboyant, debonair personality? I might add, that his parents are Catholic. Personally I am right there in Dad's corner. The decisions you make are quite restricted by your personality. You still have a choice, a free will, but it doesn't cover the whole gamut—just a subset, dictated by your character genes."

"Amen," I said. "Any more questions? I will allow one more and that's it—this theory is starting to get on my...nerves."

"OK," Andrew said,

"WHAT ABOUT...ACCOUNTABILITY?"

"Accountability?" I asked.

"If all these character traits are dictated by genes, is...anyone responsible for his actions?"

"Very good question," I replied. "But before we get into it in some depth, it is a good idea to review our present state of behavioral thinking. Do we now have accountability? Ignoring GSP, do we presently consider people responsible for their actions?"

"Well, when their actions are good, we certainly do," Estelle quickly pitched in. "We give them credit, we praise them, we hold them out as role models for emulation."

"Precisely. But what if their actions are inappropriate or downright evil—detrimental to society? Do we readily hold them accountable?" Then facing all of them, "Do you personally accept the blame for some of your questionable actions?"

Silence. Glancing around by everyone.

"I can't comment on my own behavior," Estelle volunteered, "but I have observed that very few people take responsibility for their missteps, and often when they show remorse, they do so hypocritically. Even when they are genuinely sorry for the outcome of their actions, they are convinced that some external factor was to blame."

"You know what I think?" I intervened. "I think we all have a gene which encourages us to accept credit for good acts, but compels us to reject the blame for bad ones. And this holds not only for individuals but for tribes and even whole nations."

"But if you postulate such a gene—which by the way, is not in your original list—then according to your GSP theory, this gene must be good for the evolution of the species, right?" Paul inquired, with an "I found a flaw in your theory" look in his face.

"Certainly," I readily replied, disappointing him. "We get upset about this kind of behavior—the...*blame game*, as it is often called—but it is quite important for one's self-esteem. Show me a guy who blames himself for his mistakes, and I will show you a perfect candidate for mental depression, or even suicide. In fact, I am predicting this very minute that depression and suicide may be directly linked to a defective *blame gene. A normal, human being, must find someone else to blame for his mistakes, if he is to keep his sanity.*"

"I think I see where Dad is going?" Christopher jumped in. "If we all have this *blame gene*, then all the behavioral theories we can develop are bound to be nothing but excuses for our inappropriate behavior. Right Dad?"

"Precisely." I replied. "Do you know why people are enamoured with present psychological behavioral theories?"

"I will tell you why!" I continued, answering my own rhetorical question. "Because all the explanations blame someone or something else—never the individual who performed the wrongful acts. They are not really theories of human behavior—just a collection of excuses for such behavior. The theorists are, of course, unaware of it, it is another manifestation of the...gene conspiracy, this time through the *blame genes* dictates."

"Can you be a bit more specific?" Andrew wanted to know. "Are you actually saying that all present psychological behavioral explanations absolve the evil doer of his acts?"

"Yes," I replied. "Do you know of any psychological theory that asserts that the person who did some foul act, did it because he was evil? Don't they all blame parental abuse, wrongful toilette training, societal neglect, childhood traumas, divorces, and the like? Heck, I even know a family that blamed all their problems to the fact that they were forced to move a lot—it was the nature of the father's job."

"I totally agree with Dad, here," Christopher said. "And I even know why these theories are popular, besides the fact that they give us all these excuses."

"Why?" Paul inquired.

"Because, they can not be proven wrong—that is why. You can always claim some childhood trauma for your misstep—it's hard to verify it?"

"But that has its pitfalls, too." I added. "These theories are popular with behaviorists and…armchair scientists, but when they are utilized in the justice system, they are generally dismissed. The fact that they cannot be proven, and that experts disagree among themselves on the explanations, renders them speculative. Jurors usually ignore them, or at best, they use them as mitigating circumstances."

"I'm glad that you have finally concluded discussing generalities and reached the serious issue of accountability in the justice system," Paul said. "Considering punishable offenses—not just…weird behavior—what are you really advocating? Is it not just a new excuse for our missteps? Won't *my genes did it*, become a new court room…defense—a way to commit crimes and get away with it? If that is the case, not many people will buy this GSP theory of yours, I am afraid."

"For court room drama, I like the *old the devil made me do it* excuse," Andrew intervened. "It is a bit more…believable, especially if the jury is made up *of God fearing* people."

"I got to agree that the phrase, *the genes made me do it* does not have a nice ring to it," said Christopher, "but it maybe due to a lack of familiarity with the GSP. I just read recently in an Internet article that some sleazy lawyer is developing this defense for a…pedophile. So maybe we are not too far from actually absolving ourselves of all responsibility for our criminal actions."

"Well, if this theory leads to that, so be it?" I replied. "A good example is Tourette's syndrome. We used to hold the afflicted responsible for their cussing and swearing—we are gradually accepting their inability to control it. If it turns out that our genes do dictate people's actions as much as I am contenting here, how can you hold anyone responsible? "

"Are you done with these digressions? Can we return to our main issue—the legal issue?" Paul insisted.

"Yes, certainly," I replied. "For starters, I think that in certain cases, the excuse *my genes made me do it* is already accepted, though camouflaged under different verbiage."

"You mean like during wars, when it is OK to kill the enemy?" Andrew asked.

"No, I didn't mean that—society does not only excuse killing in wars, it actually...encourages it." I replied. "We are discussing crimes under the penal code. The principle of self defense is essentially *a "my survival genes made me do it"* statement. It absolves you of responsibility even in cases when there was no...verifiable life threatening danger—just a perception of it. Essentially we have already accepted, that if the gene...is awakened, our ability to decide is...defunct—our actions are the...gene's, not ours. It is the condition we called ...genal, back in the discussion on survival genes, as you recall."

"Come to think of it," Andrew added, "We also have the legitimate defense of temporary insanity. It's been used when people have a mental problem, when they are under the influence of some drug, or if they lost their...head, in general. It is essentially saying that the brain was not functioning—the person couldn't make a conscious decision, and carry the responsibility. It is another form of the...genal condition, also. So I guess, we may soon hear that some people use GSP as a defense—especially after this book is published. We must expect it soon for the cases of defective genes, which lead to heinous crimes like rape, serial murders, pedophily and any others where a defective gene is found to be the...culprit. After that, the field gets somewhat murky. Wife or child abuse, for example, may be a tough sell, even though we sense that these defective genes are very demanding—the afflicted is unable to...arrest their dictates."

"All these things will pick up speed when the entire DNA is mapped out and all the genes are identified," I added. "Next will come a better understanding of which ones are responsible for what crimes, the effect of their strength, etc. Don't forget that a gene need not necessarily be defective to cause a crime, it may just be very strong, and violently...awaken. This could be the case with the...racism gene for instance. And by the way, it need not always be a defective gene that is causing a criminal act, it could be a brain dysfunction. The key difference will be that no *psychological* factors are invoked."

"I've got a question about this whole issue," Paul said. "Let us accept for a minute that some gene, or brain dysfunction is responsible for someone's crime, and therefore the person is acquitted. Where does this lead us? Do you put such criminals back out there on the streets, with the mathematical certainty that they will commit the same crime over again?"

"This problem already exists with rapists and pedophiles who serve out their prison terms," I said. "Some of them have admitted that they will commit the same crime over again—that they are unable to control their urge. I don't think we have a solution yet—not fully acceptable to the society. Some states have passed "notification" laws—neighbors must be informed of the existence of a pedophile in their mist. Some people advocate a return to the concept of...Devil's island—put them all in there, and let them...rape each other. I think Estelle is of this opinion, if I recall correctly."

"A final solution may exist in the future, if science finds a way to...cure defective genes, or decrease their strength." I added. "We have talked about this before on homosexuality. It is fraught with ethical type of questions. The data is not in for us to move this issue forward at this juncture."

At that point I got up and started to stretch my legs.

"Well," I said, "We have finally reached the end. Now all I have to do is gather all the notes, and put it all down in writing."

"Are you saying that we are finished here?" Christopher asked astonished. "If so what is the final conclusion? How many here are convinced that the theory is correct?"

"I am not sure," I said, "but I think we have one definite believer, you, one leaning positively the GSP way, Andrew, and two still wavering, Paul and…myself. Estelle is still an unknown—she wouldn't admit it even if she…converted. She likes to keep the…belligerence going—it keeps our marriage fresh and unpredictable."

"You don't really think we are going to buy your claim that you are…rejecting your own pronouncements, do you?" Paul asked. "This false modesty hardly becomes you."

"That is right, Dad," Andrew added. "And that nonsense about the cover-up gene, which renders us incapable of seeing the gene conspiracy? Is that another one of your gimmicks that anticipates a person's refusal to agree with you? Like saying, you are not smart enough to see my point? " Andrew asked.

"Certainly not, I said. You will see when the genes are finally mapped out." I avoided answering Paul's comment directly. I knew, that they all knew, that I knew, that Paul was right.

"There is one gene which definitely exists, and Dad has it with strength 10." Paul remarked as they were all walking away. "It is the gene that…compels you to attribute all human behavior to…genes."

I heard what he said. I also heard the laughter as they were closing the outside door. But I didn't mind. I felt a certain sense of relief—my genes in complete…equilibrium with the situation. The book was finished—I could…start on another one—another set of characters to squeeze, with my brilliant argumentation. I was in…Christopher's Paradise.

"What happened," Estelle asked some half-hour later, moving groggily towards the kitchen. "Are you finished discussing your…ESP theory?"

"It's GSP, and you…know it," I said.

"Yeah, whatever. Did you convert them to…your way of thinking?"

"You bet," I said. "You can read the entire discussion in the final version of the manuscript. I am reporting it exactly as it happened.

3

Rarely Asked Questions (RAQ's)

An Interview by Stamatis Kambanis

It has become apparent that the GSP has...far reaching implications in all fields of human endeavor. The basic book could not possibly include all these, because I couldn't squeeze them in the various discourses with much humor or buffoonery. I didn't want to ridicule some rather...sacred cows.

As I saw it, I had two choices. An interview with some imaginary journalist, or a section with Frequently Asked Questions—a quite fashionable practice nowadays because of the Internet. The title above shows that the two choices were combined into one, by a simple artificial mechanism. The title also reflects the fact that these questions are really, rarely... asked.

Mr. Stamatis Kambanis is actually a real person, an international lawyer, and an old and dear friend who kept encouraging me to write this book. The interview took place, presumably, a few months after the main book was finished.

The author

Stamatis: Where did you find the...audacity to write a book when you have no formal training in this area? A book without any references and credit to previous thinkers, cannot be considered a scholarly book, can it?

Aris: Back in my University years, I got interested in some French writers like Andre Gide, Jean-Paul Sartre and especially Albert Camus—his book, the Fall, was my...bible for decades. So, let me start with a...reference to Gide—I hope that I remember correctly that Gide wrote it—don't want him turning over in his grave.

"I wish I were a bastard" is what Gide said in one of his writing. He went on to explain this statement by saying —and here I may be taking some liberties—that as a...bastard he would not be under the...protection, and thus the...influence of his parents. He wouldn't have to think and do what they expected of him. He would be under the authority of people he disliked—free to create his own ideas about the world.

Well, I am the...bastard a la Gide, and bastards have plenty of audacity. In Formal Training, you create your ideas under the auspices of some parents who cuddle you, protect you and expect you to be...faithful—to carry the...torch. Most formally trained scientists are under system-imposed and even self-imposed constraints—content to push the field just one small step forward, within the existing framework. Don't expect much revolutionary thinking from the...establishment. They will not be happy with this book, but there isn't much they can do to...hurt my career in their...field—I don't have one.

Now references are put in to show to other colleagues who will critique the book, that you had training in the area. Every Ph.D. dissertation has a background chapter with references to previous work—an outline of the thinking up to that point. References are also used to...flatter your colleagues in the field—it is all part of the act of being a member of an select group of...learned people. You play the game, or you are ostracized. Don't forget that I have been a player myself—I was a professor of Electrical Engineering, during the peak of the...publish or perish era. Well, I don't have to play this game any more—I am retired.

Without references, the book will be judged unscholarly, and it is—no reason to pretend otherwise. But scholarly ain't all that's cracked up to be. It usually means boring and unimaginative—often just...bullshit.

By the way, the reason that I wrote this book in the style of pointed discourses is because I don't want people to take it seriously. Most of the stuff I am saying—I don't even believe myself.

With this answer I hope that I was able to convert a weakness of the book, into a...distinct advantage. Thank you for asking it!

Heavy laughter!

Stamatis: The book has been criticized for not having any historical insight. One critic said that the author, you, appears to have no knowledge of...history, and as Winston Churchill said "those who do not know history are condemned to repeat it."

Aris: First let me thank you for providing another reference for the book—we now have two in two pages.

If the GSP theory is even partially right, then, the Churchillian saying would have to be modified to "those who don't know history, as well as those who do, are condemned to repeat it."

That is...my take on GSP's application to world history. No amount of knowledge of any sort helps to prevent wars. When people perceive a threat to their existence, their survival genes are turned on with...a vengeance, logic is thrown out and all knowledge—not just historical—is...useless. There is an old saying used in journalistic circles—too bad I can't recall the author to add it as a third reference. It says that the first casualty of war is the...truth. That, too may have to be modified by replacing the word...truth, with the word...logic.

As proof of my contention that the book does not need history, I offer you...history itself. Has historical knowledge stopped any...wars? So, unless it is shown that all historical leaders involved in wars were...historical...nincompoops, like me, it doesn't seem to make much difference.

Stamatis: I took a little time to research the issue, and it looks to me that you may be a behavioral geneticist—a label given to people who advocate that behavior and personality are dictated mainly by genetics. Do you feel like one?

Aris (after a long pause): Well, if you asked me this question a month ago, I wouldn't know what you were talking about. However, having recently looked at the book "Born that Way: Genes, Behavior, Personality" by William Wright, I understand and, therefore, I can try to answer your question.

I don't think that the GSP theory is the same as Behavioral Genetics, though they may share some common elements. The GSP theory does not advocate that personality and behavior are dictated by inherited genes, but by the totality of genes as well as other chemical/electrical processes in the brain. If these change during one's lifetime—and they can, as we have often illustrated—personality will also change, and it will no longer be what the inherited genes dictated.

I think the proper label for advocates of the GSP theory might be behavioral...chemists, not geneticists.

I think the common element in both theories is that they consider psychoanalysis as nonsense—incapable of solving mental problems or changing human behavior.

Stamatis: Let me zero in on the survival genes—the violent ones in particular. You state in the section on survival that these genes wake up under a perceived threat, and totally overtake a nation—they freeze it in a...wholesale genal condition. Many feel that these are pretty strong and even...irresponsible conclusions. Do you have any proof that this is the case?

Aris: Let me see if I can come up with some—this is where historical knowledge would come in...handy, I guess—to support my arguments."

Head scratching.

"OK, let's take the United States during the Second World War. Before the Pearl Harbor attack, the common wisdom—brought about by logical thinking under dormant survival genes—was to stay out of the war. Then came the attack on Pearl Harbor. The people became instantly convinced that the US was under...threat. Nearly everyone went from the logical level of thinking, to the... genal level. From then on it was all patriotic flag-waving, enemies lurking even at home in the US, sacrifices and the like—logic gone, till the war was over.

Germany under Hitler is another example. I have been always amazed at the fact that Hitler was able to place such a large part of the population in the…genal condition, turning them into living…zombies, for as long as he did. It is hard to comprehend it when you operate in the…logical level—not under…genal control. The GSP theory's conclusion is that it was nothing…*endemic* to the German people. It can happen to you and me—unless your survival genes can not turn on for some reason.

By the way, such a…national psychosis is very scary for those few that don't have strong survival genes—or they have them defective—and can still think…logically. They are easily branded as traitors, as…giving aid and comfort to the enemy—the usual phrases dictated by the genes. Intolerance is the order of the day. Nature has worked it out that way, so that the response to the threat is immediate—no committees, no congressional hearings, no debating, and all the rest. In fact, the genes have…lobbied so successfully, that the genal condition has been legalized even in some democratic countries. It is usually *called, Marshal Law or State of Emergency* or whatever, and it is accepted as a reason for suspending all basic freedoms.

You can see how easy it is for a good…demagogue to exploit all this. With…charisma, he can convince people of some imminent threat—the usual ploy is to…create one—and put them under…genal control, which usually translates under…his control. Sometimes this…genal level hangs around for years—all he has to do is keep feeding the notion of the…threat. When he goes, either by catastrophe or natural death, the…genal level is removed and the survival genes go back to the…barracks. It is then that logic returns, and the soul searching begins—a very painful state of affairs for a nation. It can not easily admit guilt for its past—that too is threatening to its survival. It usually takes a couple of generations to be able to come to grips with a period of…genal control and its excesses—the people who lived it, cannot do it.

Aside from country leaders, some...cult leaders also use the technique successfully. In general, most of these groups are highly dependent on...charismatic leaders, who are convincing enough to bring up the...genal level. As soon as they die, the movement dies with them.

Stamatis: You paint a very bleak picture here, and the implication that these things are inevitable—that the people bear no responsibility—is even bleaker. Is there anything we can do? Can we come up with some possible solution using the principles of...GSP?

Aris: I think, for this problem, the key is to create new nations—nations that don't resemble the old tribes, for which these genes evolved.

Let me explain further. The survival genes operate optimally, when people live in small, homogeneous units. If we alter these conditions, the genes will become less effective. This is fraught with dangers, of course—you solve one problem but create a slew of others. You render some genes ineffective, but others raise their...ugly head, causing new inconveniences.

To be more specific, since the survival genes evolved to protect a small, ethnically pure, strong-cultured, cohesive unit, the farther away you get from that, the less likely the country is to drop to the...genal level, at the drop of a hat. Big, multi-racial and multi-cultural societies may turn out to be the answer.

The United States is such a country, and it is quite difficult to put it in a genal condition—not impossible, just difficult. The threat must be real—not faked—and all these ethnically diverse groups must perceive it as such. These are things that future aspiring demagogues must content with.

Europe is trying to create such a multi-national entity, but it will be difficult—it is easier to do with a new nation. It will be difficult because it is a...catch 22 situation. To decide to do it, you must operate at the logical level. But when you actually start doing it, it is perceived as a threat, and the genal level pops out and stops you. The manifestations of this catch 22, are everywhere in Europe.

Success in creating a multi-racial society ameliorates the genal condition problem, but it causes a new set of problems. The genes have not changed, and they don't like this new environment. They prefer the old homogeneous unit they have evolved for. So now you have your personality genes at worse...equilibrium than before. with the usual increases in stress, angst, etc. People start joining groups, clubs, and churches—anything resembling the small uniform groups the genes desire. This is hard to do in big cities. That is why small town living is often perceived as...happier living.

So, you see, there are no easy solutions—just a fresh approach for studying our problems. New reasons to spend millions on proposed research. The test of whether the theory leads to solutions is yet to come. Personally I am not very...optimistic. People are very complicated entities, and the...Paradise defined by Christopher during the Symposium will be hard to create.

Stamatis: I'd like to return to the problem of racism for a minute. You told Estelle that you had no solution to the problem then—you felt your duty was to mainly bring out the reason for its existence. Have you thought about the problem since the last time it was discussed?

Aris: Yes, I have, it's been some time since I wrote that episode. I might even have a possible solution—though it's hard to say at this juncture. The key lies with understanding the workings of the racist gene, so it might be some time before we know for sure. More specifically, we must unlock the mechanism of a person's ability to recognize his own kind—the average...look of his tribe. At some point in time, a newborn child creates this general image of the average ...face of his tribe, and after that any deviation from this face, alerts the person to a possible enemy—someone he must fear and dislike. Recent research, by the way, appears to bear this out.

People' s brains became agitated when they were shown photos of faces from other races—remained calm when the faces were of their own race.

But when does the person create this average tribal image? If it is before he is born—as it is with sexual images like a girl's thighs, for example—we are up a creek.

But what if the gene only dictates the creation of this average face, based on what the child...sees during his formative years? If this is the case, the answer is obvious. Racism will be eradicated when there is enough intermarriage among the races so that we all...look alike. In the meantime, a racially mixed environment will be needed from the time the child begins to see, something quite hard to do at this juncture.

You can easily surmise, that we need some additional gene research before we continue our speculations. In the meantime, we should stop blaming one another—we are all very fine racists.

Stamatis: What could be an ideal political system for survival genes? From what you have been saying, it does not seem that they are very happy with democracy, is that right?

Aris: You've got to try to imagine the political conditions for which these genes developed. I am no expert on primitive man's politics, but I doubt if democracy was a high priority in his thinking. He probably lived under continuous threats and thus under a perpetual...genal condition. If that be the case, the regime was probably a simple hierarchical structure, created anew for each generation—no hereditary aristocracy at that time. The leader evolved after some violence and machinations— he had to inspire...awe, reverence, fear and mainly faith in his ability to protect the tribe.

Democracy cannot be implemented, if the nation is in the genal level. Democracy is a regime of reason, and the nation must be free of the genal condition for it to flourish.

Stamatis: We keep talking about the traits of a leader. What is leadership, vis-à-vis the GSP theory?

Aris: Everyone talks about leadership—how important it is in a President or other type rulers—but nobody really knows how to define it. I think the key idea in leadership, is that this person must be able to inspire others to do what he wants them to do—to grab the bull by the horn and solve some problems his way. He has to have charisma, an aura about him that he is a legitimate...chief of the tribe—to look *presidential*, as it is called today. This sounds very much like a...recipe made by genes. A longing for such a leader is basically a demand by the...genes. That is why there is no mention about this leader's *position papers* on the issues of the day. Nobody gives a damn about his opinions—just the overall package.

Stamatis: While on the subject of politics, the following issue comes to mind. Do you think that some of the book's conclusions might be used by some to push their own political agendas?

Aris: I doubt it. Most people will dismiss the book as nonsense—it wasn't written by an expert in the field.

Stamatis: It was mentioned at some point in the Symposium that there are situations when two or more genes may be at...odds with each other. Can you give an example and discuss the consequences?

Aris: This type of internal conflict is usually called a *dilemma*. If you want to get married and the only girl available in your village is a...dog, you got yourself a...dilemma—the procreation gene is fighting with the attraction gene, and you have a sense of...unease, to say the least.

Such dilemmas—genes fighting with each other—can be devastating—they can lead to your destruction. Consider the following scenario. Your town is invaded by enemy troops. You immediately revert to the genal condition—logical thinking becomes...defunct. One gene wants you to fight to...save the town, another gene wants you to run to...save yourself. Either way, you are doomed—you may even get gray hair, or dirty your underwear, while the genes are fighting this out. If you run, you could be branded a...coward. If you fight, you are most likely a...dead man.

One thing is certain, though. Being in the genal condition, your decision is not...rational—you bear no responsibility for either of the two actions. The stronger of the two genes won, and you behaved according to its...dictates. But the rest of the town may not subscribe to the GSP theory—they may hang you for cowardliness, if you run—declare you a hero, if you die.

In both cases you are dead. And "when you are dead, you are dead" Curt Vonnegutt, Jr. once remarked in the introduction of one of his books—claiming it was the moral of his story. Here it represents the moral of a gene...dilemma.

Even if you survive this type of...gene struggle, your brain might suffer some chemical alteration and your personality may change from the ordeal.

Stamatis: What about suicide, especially among the young? Could that be explained in terms of genes conflicting with each other?

Aris: Well, not really. You don't have the survival genes fighting some *suicide* genes with the latter winning the fight. Suicidal tendency is probably caused by some defective gene or chemical in-balances in the brain. It is probably related to depression. I am not a expert on mental disorders. All I can tell you is that they will eventually be cured, not by any...couch technique but by some chemical drug that will correct the brain in-balances.

Stamatis: But what about teenage suicides? They don't seem to be caused by defective genes. Can they be explained by the GSP theory? What would cause young people to want to take their own life?

Aris: That's a complicated, multi-faceted issue—with, quite likely, a plethora of genes involved. I can only give you some ideas—talk around it, so to speak. Maybe even stimulate some new research.

To begin with, I am sure that girls and boys do not commit suicide for exactly the same reasons. Keeping in mind GSP, i.e. that the genes program our actions for the survival and evolution of the species, both groups are driven to it by a sense of inadequacy, a conviction that they will not fulfill this mission. For teenage girls, the first act for the fulfillment of this goal is to fall in love and mate as early as possible—don't forget that women are only fertile for 25-30 years. So they rapidly fall in love, and if the male of their affection does not reciprocate, their feelings of disappointment are monumental—they think that they have... *betrayed the species*. This enormous sense of inadequacy and failure causes the secretion of...nasty chemicals in the brain, which lead to depression and suicide. It may even be that the genes have programmed it that way—they may have no use for young girls who fail to mate at the peak of their youthfulness and productivity. The inevitable conclusion is that the preponderance of teenage girls commit suicide for romantic reasons.

For teenage boys the story is a bit different. Their first *programmed* task is to sort out the hierarchical structure among themselves. Their sense of inadequacy must be due to a realization that they rank low in the respect of their peers, and a perception that their chances for rising high are quite low. The conclusion here is that one must look towards factors involved in hierarchical climbing (school or sports failure, lack of peer acceptance, etc) when confronted with suicidal activities in teenage boys. Of course, we can not entirely preclude a "romantic" link in the reasons for the...act, but the probability is high that the romantic failure is blamed on unattractiveness due to the low rank.

Stamatis: What about the...Placebo effect? You promised Estelle to take it up and explain it using GSP, during the Interview. She asked me to remind you—she thought you might try to weasel out.

Aris: Weasel out? Is that what she said? She is developing quite an acid tongue as of late.

To briefly review it, sick people sometimes get well by taking a drug they believe can cure them—even though the drug is just a...sugar pill. When this happens, medical science attributes the cure to the... Placebo effect. It is widely believed, that the key thing behind this effect is that the patient must quite convinced that the medication will cure him.

Our pertinent question here is whether our present GSP can explain this rather weird phenomenon.

To begin with, the Placebo effect has not been explained satisfactorily by any other theory, so expecting GSP to explain it is not altogether reasonable. All the same, I will make an attempt to do so, even though the explanation might only be convincing to...myself.

Now the explanation. We know that chemicals or electrical waves in the brain can cause feelings and thoughts—we have discussed this in detail when we argued the non-existence of the psyche. We also know that the converse is true—feelings or thoughts can cause chemicals[13]. So it is not totally unreasonable that a feeling of complete faith in the sugar pill, or in the powers of a healer, can cause the production of a chemical that fights and cures some disease. Even tumors are known to have vanished under the power of chemicals produced by strong feelings of faith.

I have already mentioned in the <u>Are you...Mental</u> section that the Placebo effect is what keeps all the psychologists, healers, producers of herb medicines and other folk remedies going strong, and getting stronger. There is always some anecdotal evidence that they are effective—

---

[13] Stomach acid in stress, adrenaline when in competition, perspiration when in fear, etc.

someone is getting cured due to the Placebo effect. It has turned out to be a bonanza for quacks throughout the ages, and it will never stop.

By the way, I think that the Placebo effect is also the explanation for…miracles. It is no coincidence that Jesus Christ always demanded that the afflicted proclaim his unwavering belief in him and in his power to heal him. So do all present day miracle healers. If the person believes that you can cure him, then your chances of doing so are increased—they become the percentages of success of the Placebo effect. That is why the cures by these self-proclaimed healers/evangelists are not all fake. Some people actually get cured—*through* the evangelist—by the Placebo effect.

Incidentally, there has never being a miracle where a guy walks in with one leg, and walks out with two—the second one, grown by the miracle. You can not fake such a miracle, and the Placebo has never managed this type of cure. Even Jesus Christ did not attempt to re-grow a missing limb—or even a tiny finger. If he'd been successful in doing it, he might have gained another follower—myself.

Stamatis: One last question. You often joke about the GSP theory, sometimes even claiming that you are not sure whether you actually believe it or not. Other times you appear dead serious and quite passionate about its conclusions, yet, almost with the same breath, you spoil the mood by asking the reader to consider the whole issue as…tongue in cheek. What is the true state of your belief in the GSP theory? And if you are still wavering, how do you expect others to accept it?

Aris: To tell you the truth, I have never fully believed in my theory, I had no business developing it—no serious credentials in the field. But the real reason is, I am a…pathological skeptic—a defective gene, no doubt. And I haven't been able to shake this, even for my own theory. That is how strong the genes are—how effectively they control us.

As for the contention that my wavering will make it hard for others to accept the GSP, I am in complete agreement. Don't forget that I am not proposing this theory, the same way one would propose a...new religion. I am putting it out there more as a...*prediction*—I am essentially formulating what science is on the verge of proving in the very near future. If I am right, I don't need to convince anyone at this stage. The convincing will occur, all on its own.

The End

About the Author

The author is a semi-retired professor of electrical engineering. He has authored, co-authored or edited many engineering books in English and other languages. He has also co-authored two books on Backgammon, and dozens of scientific articles in his research fields of Communications, Signal and Image Processing and Radar Systems. All his previous books have been written under his real name, Nicolaos S. Tzannes. Aris P. D'Avenal is a pseudonym, which the author is using here for the first time.

The author is quick to confess that he has no academic credentials in any of the fields the theory inevitably led him to. He does have a Ph.D.—but it is in Electrical Engineering. He has no formal training in psychology—just a healthy…contempt for it. His knowledge about the other subjects that the theory entangles (human behavior, molecular biology, history, politics, religion, etc.) come from books, articles, and, mostly, from a lifetime's keen interest in observing the human condition. For this reason, he anticipates a large controversy about the conclusions of the theory. He expects no praises from present day behaviorists—just their wrath. Not that he can really blame them. There are days when he, himself, finds the theory rather absurd—though a bit less absurd than most existing theories on human behavior. But with, or without objections, the theory is out there now, nobody can put it back in the bottle again. Its postulates lead to its conclusions, and these conclusions do not seem altogether unreasonable—no matter how hard they are to swallow. In any event, the human genome project (the mapping of all human genes) is moving right along. It shouldn't be too long before we know whether the theory deserves high praise, or the dustbin of history.

www.ingramcontent.com/pod-product-compliance
Lightning Source LLC
Chambersburg PA
CBHW061306280526
45784CB00002B/920